BUILDING FOR ENERGY
CONSERVATION

BUILDING FOR ENERGY CONSERVATION

by

P. W. O'CALLAGHAN

BSc, MSc, PhD, CEng, MIMechE

School of Mechanical Engineering, Cranfield Institute of Technology

PERGAMON PRESS

OXFORD · NEW YORK · TORONTO · SYDNEY · PARIS · FRANKFURT

U.K.	Pergamon Press Ltd., Headington Hill Hall, Oxford OX3 0BW, England
U.S.A.	Pergamon Press Inc., Maxwell House, Fairview Park, Elmsford, New York 10523, U.S.A.
CANADA	Pergamon of Canada Ltd., 75 The East Mall, Toronto, Ontario, Canada
AUSTRALIA	Pergamon Press (Aust.) Pty. Ltd., 19a Boundary Street, Rushcutters Bay, N.S.W. 2011, Australia
FRANCE	Pergamon Press SARL, 24 rue des Ecoles, 75240 Paris, Cedex 05, France
FEDERAL REPUBLIC OF GERMANY	Pergamon Press GmbH, 6242 Kronberg-Taunus, Pferdstrasse 1, Frankfurt, Federal Republic of Germany

First edition 1978

British Library Cataloguing in Publication Data

O'Callaghan, P W
Building for energy conservation.
1. Heating   2. Energy conservation
I. Title
693.8'32   TH7226   77-30332

ISBN 0-08-022120-3

Printed Offset Litho in Great Britain by Cox & Wyman Ltd, Fakenham, Norfolk

To
Menna
and
Catrin

Contents

Chapter 3   THERMAL COMFORT

Chapter 4   CLIMATE

Chapter 8   ENERGY THRIFT

Chapter 9   SECONDARY EFFECTS

Chapter 10   WASTE HEAT RECOVERY

Chapter 11   ALTERNATIVE ENERGY SOURCES

Nomenclature

a	coefficient	
b	coefficient	
c	specific heat	$\text{J kg}^{-1}\,\text{K}^{-1}$
c_d	drag coefficient	
c_f	skin friction coefficient	
d'	declination	rad
f	friction factor	
$f(\ )$	function of ( )	
f	frequency	c s^{-1}
g	acceleration due to gravity ($= 9.81$)	m s^{-2}
h	partial heat transfer coefficient	$\text{W m}^{-2}\,\text{K}^{-1}$
h'	hour angle	rad
h^*	mass transfer coefficient	m s^{-1}
i'	incident angle	rad
'j'	Colburn 'j' factor ($= St\,Pr^{2/3}$)	
k	thermal conductivity	$\text{W m}^{-1}\,\text{K}^{-1}$
k^*	permeability	$\text{kg m N}^{-1}\,\text{s}^{-1}$
l'	latitude	rad
m	mass	kg
$\dot{m}$	rate of mass flow	kg s^{-1}
n	index	
n'	solar wall azimuth angle	rad
p	pressure	N m^{-2}
$\dot{q}$	rate of heat flow per unit area	W m^{-2}
r	radius	m
s	pitch	m
t	time interval	s
t°	sun-time	s
u	velocity	m s^{-1}
w	skin wettedness factor	
x	abscissa	
x'	fractional quantity	
y	ordinate	
z'	azimuth angle	rad
A	area	m^2
$\mathscr{A}$	absorptivity	
B	grouped parameter ($= \Delta x^2/\alpha\Delta t$)	
Bi	Biot modulus ($= h\,L/k_s$)	
C	thermal conductance	W K^{-1}
C^*	permeance	$\text{kg N}^{-1}\,\text{s}^{-1}$
COP	coefficient of performance	

D	diameter	m
D^*	diffusion coefficient	$m^2\,s^{-1}$
E	amount of energy	J
$\dot{E}$	rate of energy flow	W
ET^*	new ASHRAE effective temperature	°C
F	shape factor	
F^*	correction factor	
Fo	Fourier modulus $(= \alpha t/L^3)$	
$\mathscr{F}$	configuration factor	
$\dot{G}$	mass flow rate per unit area	$kg\,m^{-2}\,s^{-1}$
Gr	Grashof number $(= \rho^2 g\beta\Delta TL^3/\mu^2)$	
H	specific enthalpy	$J\,kg^{-1}$
I	radiative flux	$W\,m^{-2}$
K	constant	
L	length	m
M	grouped parameter $(= mc_p)$	
NTU	number of transfer units	
Nu	Nusselt number $(= hL/k_f)$	
Pe	Peclet number $(= Re\,Pr)$	
Pr	Prandtl number (ν/α)	
Q	heat content	J
$\dot{Q}$	rate of heat flow	W
R	thermal resistance	$K\,W^{-1}$
	(thermal resistance of unit area	$m^2\,K\,W^{-1})$
Re	Reynolds number $(= \rho uL/\mu)$	
R_v^*	vapour resistance	$N\,s\,kg^{-1}$
$\mathscr{R}$	reflectivity	
R	characteristic gas constant	$J\,kg^{-1}\,K^{-1}$
S	solar gain factor	
St	Stanton number $(= Nu/Re\,Pr)$	
T	temperature	K, °C
$\mathscr{T}$	transmissivity	
U	overall heat transfer coefficient	$W\,m^{-2}\,K^{-1}$
V	volume	m^3
$\dot{V}$	volumetric rate of flow	$m^3\,s^{-1}$
$\dot{W}$	rate of performing work	W

Greek symbols

α	thermal diffusivity	$m^2\,s^{-1}$
	altitude angle	rad
β	coefficient of volumetric expansion	K^{-1}
γ	specific humidity	$kg/kg_{\text{dry air}}$
δ	thickness	m
∂	decrement factor	
ε	emissivity	
ε_h	eddy diffusivity	$m^2\,s^{-1}$
ε_m	eddy viscosity	$m^2\,s^{-1}$

η	efficiency	
θ	arbitrary angle	rad
κ	diffusion resistance factor	
λ	wavelength	m
μ	coefficient of dynamic viscosity	kg m^{-1} s^{-1}
ν	kinematic viscosity	m^2 s^{-1}
ξ	fin effectiveness factor	
ρ	density	kg m^{-3}
σ	Stefan–Boltzmann constant ($= 5.67 \times 10^{-8}$)	W m^{-2} K^{-4}
τ	shear stress	N m^{-2}
ϕ	relative humidity	
ψ	arbitrary angle	rad
ω	heat exchanger effectiveness	
Γ	transmittance	
Δ	difference between, change in (as prefix)	
Θ	time constant	s
Φ	time lag	s

Other minor symbol usage is explained in the text.

Subscripts

a	for air		l	loss
ab	at absorber		lam	laminar
b	bulk		lt	latent
bl	black		m	by momentum
bs	base		max	maximum
c	by convection		min	minimum
cg	ceiling		n	index
cl	clothing		o	outside
cs	cold side		op	operative
d	drag		oh	humid operative
db	dry bulb		p	at constant pressure
dp	dew point		r	by radiation
eff	effective		rf	at roof
ev	by evaporation		rot	rotational
f	of fluid		s	of solid
fg	for latent heat		sa	sol-air
g	generated		sat	saturated
gd	ground		sb	sensible
gz	for glazing		sc	scattered (diffuse)
h	for heat transfer		sf	at surface
hd	hydraulic diameter		sk	of skin
hs	hot side		st	stored
i	inside (index)		sub	subjective
irf	inside roof		t	at time t
j	at stations $j = 1, n$		tb	turbulent

th	thermal	*H*	horizontal
tot	total	*I*	inlet
tr	by transmission	*L*	at position *L*
v	for vapour	*LL*	longitudinal
w	at a wall	*M*	by metabolism
wb	wet bulb	*N*	north
x	at position *x*	*O*	outlet
y	at position *y*	*P*	of perimeter
∞	at an effectively infinite distance from	*S*	south
		T	transverse
1,2	at stations 1,2, etc.	*V*	vertical
12	from station 1 to station 2, etc	*W*	west
AV	average value	δ	direct
D	based upon diameter	λ	at wavelength λ
E	east		

Other subscript usage is explained in the text.

Chapter 1

Introduction

1.1　THE NEED TO CONSERVE ENERGY

The hopes and aspirations of civilisation were disturbed during the late 1950s when the occasion of nuclear war appeared an inevitable and imminent probability. A second setback was experienced 10 years later when it became generally apparent that, if prevalent rates of increasing energy usage were allowed to continue, the world's total fossil fuel reserves would become completely depleted within our children's lifetime. Even if the annual rate of consumption were to remain constant at 1975 levels, the diminishing availability of fuel would result in debilitating shortages which would evoke drastic changes in economic policies and sociological behaviour.

Table 1.1 (derived from references [1] to [5]) presents data concerning estimated world recoverable reserves of coal, oil, natural gas, uranium, shale, and tarsand. The thermal energy content which would be released by combustion is expressed in each case in MJ. This allows estimates to be made of total "life" at current rates of consumption and facilitates comparisons. Previous "life" estimates [1, 3, 4, 6, 7] have considered each fuel alone and hence have neglected the effects of switching from one source to another as supplies weaken. Conversion constants are listed below to aid interpretation.

Energy

1 therm	$= 1.055 \times 10^2$ MJ
1 kWh	$= 3.6$ MJ
1 Btu	$= 1.055 \times 10^{-3}$ MJ
1 calorie	$= 4.184 \times 10^{-6}$ MJ
1 kcal	$= 4.184 \times 10^{-3}$ MJ
1 tce (tonne coal equivalent)	$= 2.88 \times 10^4$ MJ
1 toe (tonne oil equivalent)	$= 4.54 \times 10^4$ MJ
1 m^3 natural gas	$= 0.384 \times 10^2$ MJ
a barrel oil	$= 6.3 \times 10^3$ MJ
1 tonne brown coal	$= 8.0 \times 10^3$ MJ
1 tonne uranium metal	$= 8.1 \times 10^{10}$ MJ

Power

1 h.p.	$= 7.45 \times 10^{-1}$ kW
1 Btu h^{-1}	$= 0.29 \times 10^{-3}$ kW
1 kcal h^{-1}	$= 1.162 \times 10^{-3}$ kW

1

TABLE 1.1. WORLD RECOVERABLE ENERGY RESERVES (IN 1974)

	Coal	Oil	Natural gas	Others[a]	Total	Reference
Estimated total world resources (10^{15} MJ equiv.)						
Highest estimates	138.0	16.0	13.0	1.7	168.7	1
Average estimates	85.0	13.0	8.0	0.6	106.6	1
Lowest estimates	32.0	11.0	4.0	0.5	47.5	1
World resources known to be extractable (10^{15} MJ equiv.)						
Highest estimates	63.0	4.0	1.8	0.7	69.5	1
Average estimates	32.0	3.8	1.5	0.6	37.9	1
Lowest estimates	3.7	3.6	1.3	0.5	9.1	1,2
Annual rate of consumption[b] (1974) (10^{15} MJ equiv.)	0.086 (30%)	0.129 (45%)	0.055 (19%)	0.017 (6%)	0.287 (100%)	1,2,4
Equivalent continuous power rating (10^7 kW)	38.9	58.3	24.6	7.8	129.6	
Estimate of life (years) based on 1974 consumption						
Highest estimate	1600	123	238	98	586	1,3
Average estimate	677	65	87	35	250	1,3,4
Lowest estimate	43	28	24	30	32	1,3
Estimate of life (years) based on a 5% growth in rate of usage						
Highest estimate	88	40	53	34	65	1
Average estimate	72	26	32	21	52	1
Lowest estimate	25	18	15	19	20	1

[a] Others include uranium, shale, tarsand and other minor fossil-fuel resources. The life of uranium could be extended a hundred-fold with the development of successful safe fast-breeder reactors [8].
[b] World population: 3.8×10^9 at 1974 [4].

N.B. 1 kW expended continuously for   1 h uses 1 kWh   = 3.6 MJ
1 day uses 24 kWh   = 86.4 MJ
1 week uses 168 kWh = 605 MJ
1 year uses 8736 kWh = 31 449 MJ
1 MJ per year = 3.17×10^{-5} kW

The great uncertainty associated with estimates of world resources arises from political, geological, and geographical factors. "Estimated total world resources" are estimates of recoverable reserves allowing for improvements in extraction technology and price rises [1]. At best, current world fossil-fuel resources contain a convertible energy equivalent of $\sim 170 \times 10^{15}$ MJ, whilst the lowest applicable figure is only $\sim 9 \times 10^{15}$ MJ. Approximately 90% of this energy is contained in coal deposits. Data relating to energy usage are more systematic: 0.287×10^{15} MJ of this "fossilised" energy was released in 1974. This is equivalent to a continuous power rating of 129.6×10^7 kW; or 0.3 kW for each man, woman, and child living on the earth. The United States is responsible for the greatest *per capita* consumption of energy (~ 7.2 kW in 1974 [4]), whilst underdeveloped and developing countries consume relatively little (i.e. India 0.08 kW in 1974 [4]). Table 1.2 [8] lists current usages by major nations relative to the UK figure. If the world release of energy were to continue at a

TABLE 1.2. RELATIVE ENERGY CONSUMPTION PER ANNUM PER CAPITA
(1974 FIGURES [3, 8])

	Index	kW/capita
United States	1.40	7.20
Canada	1.08	5.66
United Kingdom	1.00	5.24
Average for industrialised nations	0.95	5.00
Belgium	0.83	4.85
Australia	0.71	3.72
West Germany	0.66	3.46
Sweden	0.63	3.30
USSR	0.58	3.04
Hungary	0.50	2.62
France	0.46	2.41
Ireland	0.33	1.73
Japan	0.25	1.31
Average for developing countries	0.09	0.50
World average	0.06	0.33
Nigeria	0.04	0.21
India	0.02	0.10

Amount of energy required by man as food for basic survival: 0.15 kW/capita.

constant rate equal to the 1974 rate of usage, complete depletion could be expected in less than 586 years, whilst lowest estimates predict that only 32 years remain to exploit recoverable fossil fuels (Table 1.1). Furthermore, annual rates of energy exploitation are not historically constant. The present world average rate increases at $\sim 5\%$ per annum (Fig. 1.1) [8]. This growth results from the necessary increased activity of developing countries to allow expansion and the accelerating rate of consumption regarded by industrialised nations as mandatory for healthy economic advancement. An analysis of UK government statistics shows that the premise implied by the latter is unfounded: the growth rate of genuinely useful energy consumption since 1900 has been less than 0.5% per year whilst the rate of prime energy conversion, due mainly to the generation of electricity, has been very much higher (of the order of 5% in the United Kingdom [9]). The "upturn" in world economic affairs expected to occur during the 1980s can only be short-lived. The scarcity of fossil fuels opposes any major or longer-term revival; this would require greatly accelerated progress in technologies associated with energy production from sources other than fossil combustibles. If the current rate of growth of energy consumption were allowed to continue unabated, then the time available to develop the necessary production plants is limited to between 20 and 65 years, neglecting the "telescopic" effects of artificial shortages created by international pricing and rationing which will undoubtedly ensue (Table 1.1).

A significant correlation exists between the rate of energy transformation and gross national product (GNP) for different countries [6] i.e.

$$\text{GNP} (\pounds/capita \text{ year}) = 7.5 \times 10^{-4} \times (\text{energy consumption } (kJ/capita \text{ year})$$

Recently, the United States has diverged from this law, using a greater amount of energy per unit GNP than the above relationship suggests. Gross national product is regarded as a major

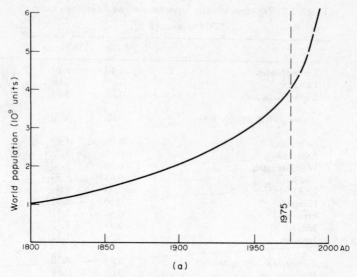

(a)

Fig. 1.1. World population growth and rates of energy consumption: – – – projections.

factor contributing to a nation's standard of living, and so it is certain that underdeveloped countries will seek increasing rates of energy production in order to raise living standards to the levels associated with industrialised areas. Internal economies in these countries (using typically 0.5 kW/*capita* at present) are not feasible and so most cutbacks must be borne by the major developed nations.

Rates of energy usage may be decreased by (a) a reduction in population, (b) a reduction in demand, or (c) a reduction in consumption.

Demographic growth in the United Kingdom is relatively constant (less than 0.5% per annum [8], whilst the United States and the USSR maintain approximately 1%; the

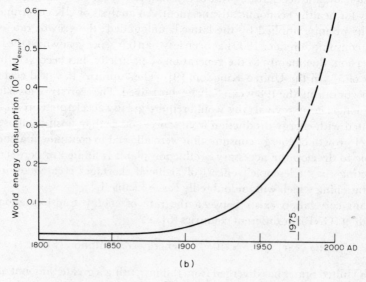

(b)

Fig. 1.1b

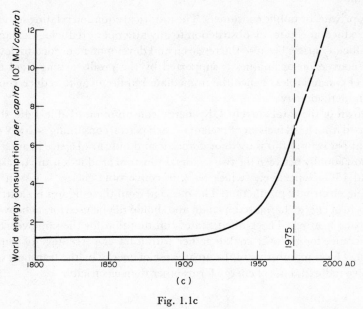

Fig. 1.1c

average rate for developing countries is $\sim 2.5\%$. Significant reductions in world population, in the absence of serious catastrophes, is not envisaged. The demand for energy may diminish by altering patterns of marketing and consumer spending. Ecologists already protest against the manufacture of energy-intensive products such as non-returnables (15 % annual increased production [6]), artificial fertilisers (10 % annual increase [6]), aerosol containers, synthetic fibres (7 % annual increase [6]), packaging, and other "labour-saving" or "convenience" products. The rate of production of non-essential goods and services must be reduced in order to conserve not only energy but also the materials required for future power plants. It is not expected that living standards would fall dramatically if some of these items were eliminated. If the world rate of energy consumption returned to the level predominant in 1950 (Table 1.3) the total average estimated life of fossil-fuel reserves would quadruple, whilst reduction to 1900 standards would result in a tenfold increase. Unfortunately, as Fig. 1.1 shows, instead of reducing demand, or even maintaining steady rates of consumption, the world is, as a whole, accelerating towards energy bankruptcy. There is no significant evidence of thrift

TABLE 1.3. ESTIMATED LIVES OF ENERGY RESOURCES

Consumption per year (10^{15} MJ)	Rate at indicated level	Average estimates		Lowest estimates	
		Life (years)	Date of depletion	Life (years)	Date of depletion
0.287	1974 level	250	2224 AD	32	2006 AD
—	5 % growth	52	2026 AD	20	1994 AD
0.220	1970 level	326	2300 AD	42	2015 AD
0.129	1960 level	556	2530 AD	71	2045 AD
0.075	1950 level	956	2930 AD	122	2096 AD
0.029	1900 level	2474	4448 AD	316	2290 AD
0.010	1800 level	7175	9149 AD	918	2892 AD

Reference sources: [1, 5, 6, 7, 10].

by personal, private, or public consumers. The historical equation relating standards of living to energy production creates an obstruction to any attempt at reducing demand for energy. A further difficulty arises because the research and development needed to develop alternative energy conversion techniques is supported by the profits arising from the profligate production of consumables: hence the immediate requirement to reduce consumption by conservation methodology.

A breakdown of the total current UK energy consumption (Table 1.4a) shows that the energy released from fossil fuels is equivalent to each person consuming ∼ 5 kW continuously. Twenty-eight per cent of this is used for domestic and industrial space heating (Table 1.5a) divided almost equally between the two sectors. Industrial production and transport accounts for 16% and 15% respectively, whilst 30% of convertible energy is lost in conversion— mainly during electricity production. The message from these figures is clear. Electricity is the purest form of energy harnessed by man and should not be used for such low-grade applications as space heating. The United States consumption for the same year (Table 1.4b) shows a preference for gas over coal, together with a lower percentage of conversion losses (Table 1.5b). This is mainly due to the greater use of energy by the transport sector and the more intensive industrial use of energy forms other than electricity.

TABLE 1.4a UNITED KINGDOM ENERGY CONSUMPTION (1974)

	Coal	Oil	Natural gas	Others	Total	Reference
Annual Consumption (10^{12} MJ)	3.20 (36%)	4.30 (48%)	1.06 (12%)	0.25 (2.8%)	8.81 (100%)	2, 3, 10
Percentage of world consumption	3.7%	3.3%	1.9%	1.4%	3.1%	
Per capita consumption[a] (10^4 MJ/*capita*)	5.72	7.7	1.9	0.45	15.75	2, 8
Equivalent total continuous power rating (10^7 kW)	10.13	13.6	3.36	0.79	27.90	
Equivalent continuous power rating *per capita* (kW/*capita*)	2.07	2.43	0.60	0.14	5.24	2, 3, 6, 8, 10

[a] Population 56×10^6. Number of households 19×10^6.

TABLE 1.4b. UNITED STATES ENERGY CONSUMPTION (1974)

	Coal	Oil	Gas	Others	Total	Reference
Annual consumption (10^{12} MJ)	10.8 (16.8%)	31.6 (47.3%)	17.6 (26.3%)	7.0 (10.5%)	67.0 (100%)	2, 3, 4
Percentage of World consumption	3.7%	10.9%	6.0%	2.4%	23%	
Per capita consumption (10^4 MJ/*capita*)	3.6	10.5	5.9	2.3	22.3	2, 8
Equivalent total continuous power rating (10^7 kW)	33.8	100.3	55.7	22.2	212	
Equivalent continuous power rating *per capita* (kA/*capita*)	1.15	3.40	1.89	0.76	7.2	2, 3, 4

Population 300×10^6.

TABLE 1.5a. BREAKDOWN OF ENERGY CONSUMPTION IN THE UNITED KINGDOM

			1974 usage (10^{12} MJ)	
Conversion losses (mainly in electricity generation)[a]	30%	30%	2.64	2.64
Industry				
Heating of buildings	14%		1.23	
Other use	16%		1.41	
Total		30%		2.64
Domestic				
Heating of buildings	14%		1.23	
Other use	4%		0.35	
Total		18%		1.58
Transport	15%	15%	1.32	1.32
Others	7%	7%	0.63	0.63
Total		100%		8.81

[a] In 1974 the generation of electricity accounted for 32% of all primary fuel usage. Average generating efficiency was 29% resulting in a loss of 23% of the energy equivalent of the primary fuel. A further 2.3% was lost in transmission.
References [2, 10, 11]

TABLE 1.5b. BREAKDOWN OF ENERGY CONSUMPTION IN THE UNITED STATES (DEDUCED FROM [3])

			1974 usage (10^{12} MJ)	
Conversion losses (mainly in electricity generation)	16%	16%	10.7	10.7
Industry				
Heating of buildings	9%		6.0	
Other use	23%		15.4	
		32%		21.4
Domestic				
Heating of buildings	9%		6.0	
Other use	6%		4.0	
		15%		10.0
Transport	25%	25%	16.8	16.8
Others	12%	12%	8.1	8.1
Total		100%		67.0

1.2 HOW MAY ENERGY BE SAVED?

The four major energy-consuming sectors are: electricity generation, transportation, production, and the heating of buildings. Although present considerations are primarily limited to the consumption of energy by buildings, the effects of applying conservation measures in each of these four sectors will be briefly examined.

1.2.1 **Electricity generation**

It has already been demonstrated that the generation of electricity is an extremely wasteful process in terms of energy conversion. Thus the use of electricity for purposes other than those which indisputably justify the use of such a high-grade form of energy should be positively discouraged. Electricity prices per useful MJ can be up to six times that for oil, coal, or gas. Whilst generating boards strive to obtain maximum overall fuel savings, the work process involved is limited by thermodynamic principles to have a low intrinsic thermal efficiency. Even the most modern power station in operation converts only $\sim 36\%$ of the energy content of the primary fuel to electricity. The overall average efficiency for electricity production in the United Kingdom is 29%, the remaining 71% is discharged into rivers and estuaries as the sensible heat contained in warmed water at an average temperature of 70°C. It is unlikely that major improvements in the transformation process will be achieved in the short term [12] since any worth-while measures will be based on alternative fuel sources or will require radical changes in the science and technology involved. The greatest short- to medium-term economies may be gained by using the waste heat for district heating or other purposes.

District heating using the waste heat from power stations is limited to areas a few kilometers distant from the generating sites. Nevertheless, if installations were established, some of the 71% of energy wasted could be utilised (possibly up to 60% [12], representing a 12% saving in total UK energy expenditure).

The overall efficiency of the electricity production process might be increased by:

(a) The development of safe fast-breeder nuclear reactors. These could produce three times our present total energy requirement in the long term using the same annual quantity of uranium fuel as is used in existing nuclear installations [4, 8, 13], and so could supply the world's total energy needs for at least 100 years corresponding to the existing reserves of uranium fuel [13].

(b) The more extensive application of combined cycle generation. This system, by doubling the efficiency of energy conversion, could save 15% on the nation's fuel bill in the moderate term (i.e. after 1990).

(c) The use of thermionic, thermoelectric, or magnetohydrodynamic generators. These require rather longer-term development but estimates [12] have shown that 14% of total fuel could be saved by these means from 2000 AD onwards.

Survival in the longer term depends on the utilisation of inexhaustible sources of power:

(a) Nuclear fusion reactors can in theory provide all our energy needs for an effectively limitless period (10^{10} years [13]).

(b) Approximately 0.5% of the electricity generated in the United Kingdom stems from hydroelectric sources. Although the world's current energy requirements could be sated by the development of all possible sites of hydroelectric generation, the United Kingdom is disproportionately endowed with such sites and at best could only double its present rate of water-power exploitation [12].

(c) The Severn estuary presents the most probable site where electricity might be generated using tidal power. A strategic barrage could produce 1.5% [12] of the total UK energy demand until silting or ecological diseconomy ensues [8].

(d) If the power in the waves around the entire British coastline were harnessed, 22% of the total UK energy needs would be provided.

(e) The harnessing of other low-grade energy sources (solar, wind, geothermal energy), whilst not considered to be capable of contributing appreciably to the total energy required for electricity generation [12], may offset the demand for low-grade energy if these sources were used for diffuse small-scale local applications (i.e. the direct heating of buildings).

1.2.2 **Transportation**

Energy economics are riddled with anomalies. One such irregularity arises from the incompatibility of trends in industrial production and transport provisions. The economies of scale have created vast industrial conurbations which require very wide suburban catchment areas. Current town and country planning rationale separates residential and industrial regions. The private motor-car has thus become an absolute necessity to many workers, and yet its use is, and must be, discouraged in favour of public transport. But public transport is being restricted, and rising costs are beginning to erode its economic advantage over personal transportation systems. There is a complete absence of an adequate alternative means of mass transportation to the private car for trips of less than 10 km (over 70% of all mileage covered in the United Kingdom [11]). Any major energy savings in the transport sector behoves a complete reappraisal of suburban and inter-city transport planning and a redistribution of centres of population with respect to workplaces.

Although traction engines are notoriously inefficient (spark ignition 25%, diesel 35%, gas turbine 35%, stirling 45%), the energy wastage is small ($\sim 7\%$) compared to losses in other energy-consuming sectors. It has been shown that if one-third of passengers shifted from private car to public transport, only 0.3% of the total fuel used in the United Kingdom would be saved. Other suggested economies, such as the use of smaller vehicles, the development of improved engines, or a switch to diesel-fuelled vehicles, produce similar low figures (3.3%, 3%, or 3% respectively). The reduction in speed limits enforced by statute in 1974 has been estimated to have resulted in only a 0.1% to 0.5% saving in overall fuel costs. The motorist seems to have suffered much during the energy crisis whilst much energy wastage occurs elsewhere [6, 7, 12].

1.2.3 **Industry**

Industrialists are continuously striving to improve the overall efficiencies of production techniques from which manufacturing profits accrue. It follows that very few extra savings could be achieved unless prevalent economic tactics alter considerably. The commitment to a high level of employment opposes increasing trends for energy-intensive production techniques and automated systems. A return to labour-intensive operation would save up to 10% of the total nation's primary fuel commitment. Another fuel-saving possibility lies in the more widespread practice of reclaiming scrap and waste. The production of metal from scrap requires significantly less energy than its conversion from basic ores. The problems are associated with collection and sorting rather than technical transformation processes. Major savings could be accomplished by critically examining the heating requirements of industrial and commercial buildings. Often, however, managers are so involved in production that energy requirements, particularly for space heating, become of very minor importance.

1.2.4 **Heating of buildings**

Previous estimates [7, 14] have shown that up to 50% of the energy used to heat buildings could be saved by the application of:

increased levels of thermal insulation
draughtproofing

slightly lower room temperatures
optimum ventilation requirements
optimum building design
optimum building balance points
more efficient air- and water-distribution systems
alternative energy sources (solar, wind, waste heat, or waste material incineration)
improved burner efficiencies
the more effective use of clothing
heat pumping
improved lighting designs and more efficient luminaires
heat recovery from exhaust air, water, and lighting
energy auditing
the use of district heating
integrated design procedures

Energy flows should be redirected, inhibited, and enhanced in order to perform tasks with maximum efficiency. If a 50% reduction in heating requirements could be achieved world-wide, the extra time "bought" for research and development to avoid permanent global economic incapacity would be 44 years (based upon a constant 1974 energy usage reduced appropriately) or 9 years if a 5% growth rate were maintained. Energy conservation, especially in space conditioning, represents the only convenient short-term partial solution to energy flow problems which will not enforce a depreciation in living standards. The reduction of energy consumption in buildings, although a palliative measure, is essential to avoid radical changes in life-style rationale resulting in technological civilisations—based upon economic growth and accelerating environmental exploitation—coming to an end.

It is interesting to note that this 50% reduction would release enough primary fuel to support the entire transport sector. The savings incurred by a rejection of central electricity generation in favour of localised conversions with direct waste heat employment could supply *all* industrial energy demands.

1.3 THERMAL ENERGY CONSERVATION IN BUILDINGS

This book is intended to encourage the application of energy thrift by consumers in industry and in the domestic and commercial sectors by describing the techniques of conservation technology and by the use of simplifying assumptions and commonplace analytical procedures.

The instantaneous heating (or cooling) load required to maintain an internal environment at a steady predetermined temperature depends upon the rates and directions of heat transfers across the enclosing walls. External transmission modes are by convection to ambient air and radiation to adjacent surfaces and to the sky. The overall heat transfer coefficient is a transfer function whose magnitude depends upon the inherent insulating properties of the system boundaries, radiative heat transfers, and surface boundary-layer effects. Established procedures for evaluating optimum internal design temperatures and humidities for human thermal comfort are presented together with climatic data. Steady-state and transient heat transmission rates are computed for typical components of building structures, and effective heating or cooling requirements associated with solar gains and internal sensible and latent heat generation are considered.

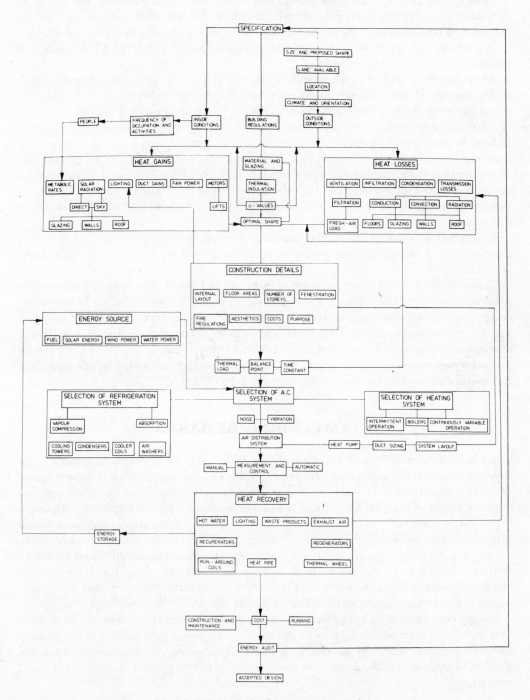

Fig. 1.2. Building design for energy conservation.

Thermal energy flow paths through building assemblies are multivarious. The reduction of systems to analogous electrical networks, in which the heat flow through each circuit element can be considered as being one-dimensional, allows complex multicomponent units to be examined in detail. Possible conservation effects may then be critically examined in relation to the whole thermal system.

When lagging is applied at the boundary of a heat container, the immediate useful result is a reduction of internal energy requirements for space conditioning. Unfortunately, secondary undesirable effects inevitably occur. A building, insulated to reduce winter heating costs, may present a formidable cooling load in summer—even in temperate climates. Indeed, the annual overall energy requirement may become greater. Similarly, the primary fossil-fuel transformation involved in producing, installing, and maintaining an insulant may be greater than the total savings accrued during its lifetime. Furthermore, the more complicated an insulated system becomes, the more moisture and dirt traps are formed. Condensation and corrosion problems are rife in lagged structures and often result in severe structural damage.

A building is a complicated and intricate thermal assembly. Partial energy-saving exercises should not be attempted without analysing the effects on the system as a whole (Fig. 1.2).

Design principles and procedures for exhaust air and water heat recovery devices are illustrated, and a review of rising new technologies, including the utilisation of alternative energy sources, self-adjusting thermal barriers, and thermal accumulators, is provided. Finally, energy management is discussed and software procedures for the examination of energy-consuming systems and the elimination of energy wastage are presented. Future prospects in energy engineering are assessed. Much of the data necessary for the analysis of the most commonly encountered thermal systems are included.

1.4 CONCLUDING REMARKS

The effects of the depletion of world energy resources have already been demonstrated by supply deficiencies arising from international trading factors. Recent price rises have shown that fuel costs permeate the whole economy. Thus, as the availability of fossil fuels becomes strained, standard of living indices will fall proportionately. The mathematics of depletion reveal that a "take-off" point is reached at the half-life of commodity reserves, after which prices accelerate exponentially unless alternatives can be found. This situation is expected to occur for fossil fuels between 2000 AD and 2100 AD, although international exchange restrictions will tend to bias the event to occur nearer the earlier date. It will then not be feasible to continue without drastic and unpopular conservation measures and a complete reappraisal of economic policies [15]. During the interim period *conservation buys time*.

Energy conservation techniques are capital-intensive. Currently, overheads are so high that most companies and individuals cannot afford the investment needed. The situation demands intensive action by governments in grants, subsidies, tax concessions, the introduction of inverted tariffs, and much more positive encouragement than is available at present.

Chapter 2

Fundamentals

2.1 HEAT GAINS AND LOSSES

The rate of artificial heat generation or removal $\dot{Q}$ to maintain a comfortable internal environment depends upon the thermal balance of the structure.

In general,

$$\dot{Q} = \dot{Q}_g \sim \dot{Q}_{tr} \sim \dot{Q}_a. \tag{2.1}$$

The natural internal heat generation $\dot{Q}_g$ accrues mainly from metabolic rates, process work, and lighting, whilst the net heat transfer across the structural boundary $\dot{Q}_{tr}$ depends upon the relative amounts of solar gains and transmission losses. The heat required to condition air for ventilation purposes $\dot{Q}_a$ depends upon the necessary rate of air change and the difference between the outside environmental temperature T_o and the internal temperature T_i required for human comfort. The transmission loss

$$\dot{Q}_{tr} = \sum U A \, \Delta T, \tag{2.2}$$

where U $(\mathrm{W\,m^{-2}\,K^{-1}})$ are heat transfer coefficients of the boundary components. For solar radiative gains,

$$\Delta T = (T_{sa} - T_i), \tag{2.3}$$

where T_{sa} is the outside sol-air temperature (cf. Chapter 4), whereas, for transmission losses

$$\Delta T = (T_o \sim T_i). \tag{2.4}$$

Overall heat transmission takes place by the combined modes of convection, radiation, conduction, and mass transfer. Rates of heat transfer by convection depend upon the nature of fluid flow, and so a knowledge of fluid dynamics is required to ascertain convective heat transfer coefficients.

Definitions

Skin friction coefficient c_f $\quad = \tau_{sf}/\rho u^2/2$.

Fanning friction coefficient f $\quad = \Delta p/(L/D)\,\rho(u^2/2)$

$\qquad\qquad\qquad\qquad\qquad = 4c_f$ for a pipe or duct.

Partial heat transfer coefficient h $\; = \dot{Q}\quad/A\,\Delta T$.

Coefficient of dynamic viscosity $\mu = \tau_{sf}/\Delta u/\Delta y = \tau_{sf}/(du/dy)$.

13

2.2 FLUID FLOW

By definition a fluid can offer no permanent resistance to shearing stresses. The application of a force causes a fluid to flow. Fluids are divided into liquids which are considered to be incompressible and gases which are easily compressed. A liquid presents a free surface at a boundary with a gas, whereas a gas expands to fill any space within which it is confined. When a fluid is in motion, shearing forces are set up between adjacent layers moving at different velocities. The dynamic viscosity μ of a fluid determines its ability to resist the resulting shear stresses [16].

2.2.1 Fluid properties

The fundamental transport properties of fluids are density ρ, dynamic viscosity μ, specific heat c, thermal conductivity k, and the coefficient of volumetric expansion β (see Appendix IV). For an ideal gas

$$\beta = 1/T, \tag{2.5}$$

where T is its absolute temperature. The kinematic viscosity, or momentum diffusivity ν, is defined as

$$\nu = \mu/\rho. \tag{2.6}$$

2.2.2 Types of flow

A fluid in motion consists of a very large number of submicroscopic particles moving in the general direction of flow. The velocity of any particle is a vector quantity which varies from moment to moment. The path followed by a particle is called a streamline.

Laminar flow occurs when sufficient viscous drag is present in the flow to damp down disturbances. The fluid particles then travel in layers, each gliding smoothly over its neighbours. The fluid streamlines are parallel and do not cross one another. The phenomenon of laminar flow is governed by the law of viscosity, momentum and viscous drag being exchanged from layer to layer.

Turbulent flow ensues when inertia forces are sufficient to overcome viscous drag—and so disturbances are allowed to grow. The fluid particles move in irregular paths resulting in confused streamlines. Because momentum is transferred in eddies as the particles move from one plane to another, rates of shear stress and momentum transfer in a direction perpendicular to the general direction of flow are much greater than those encountered in laminar flow. The eddy viscosity ε_m is defined by

$$\tau_{tb} = \rho(\nu + \varepsilon_m)\frac{du}{dy}, \tag{2.7}$$

where τ_{tb} is the effective shear stress produced and $\rho\varepsilon_m$ is the extra contribution to shear resistance introduced by the onset of turbulence [16].

2.2.3 The Reynolds number

This is a dimensionless number which characterises the nature of the flow, being defined as the ratio of inertia to viscous forces present:

$$Re = \frac{\rho u L}{\mu} = \frac{uL}{\nu}. \tag{2.8}$$

Regions in fluid flow where transition to turbulence occurs depends upon the magnitude of *Re*.

2.2.4 **Boundary layers**

A fluid flowing over a solid is retarded by viscous forces as the fluid "sticks" to the surface. Adjacent layers are slowed down less and less at positions successively further from the wall. The viscous retarding force is given by

$$\text{Force} = \tau A. \tag{2.9}$$

The region in the vicinity of the surface where the velocities are less than 99% of the free stream velocity u_∞ is termed the *hydrodynamic boundary layer* which has a thickness δ at any point x. (Fig. 2.1). There are basically two forces acting on the boundary layer: inertia forces (acting from left to right in Fig. 2.1) and viscous drag acting from right to left). As the thickness δ increases the ratio of viscous forces to inertia forces ($= 1/Re$) decreases. When the viscous forces become so small that they no longer appreciably damp down the turbulent

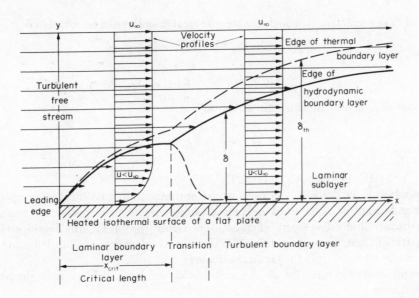

Fig. 2.1. Growth of the hydrodynamic and thermal boundary layers for uniform fluid flow over a heated isothermal flat plate.

tendencies emanating from the upstream free turbulent flow, the boundary layer itself becomes unstable and turns turbulent. In the turbulent flow regime, fluid particles adjacent to the wall are still arrested by the local powerful viscous shear, and so a laminar sublayer persists very close to the surface. The velocity in the laminar boundary layer varies parabolically from zero at the wall to 99% of the free stream velocity at the edge of the retarded region. For the turbulent section, most of the momentum of the free stream is destroyed in the laminar sublayer resulting in the characteristic velocity distribution illustrated in the figure.

2.2.4.1 FLAT PLATE

The local Reynolds number at position x from the leading edge of a flat plate is calculated from

$$Re_x = \frac{\rho u_\infty x}{\mu}.$$

When $Re_x \leq 8 \times 10^4$, laminar flow prevails.

When $Re_x \geq 5 \times 10^6$, turbulent flow dominates.

The region between these two positions is known as the transition region.

The thickness of the hydrodynamic boundary layer δ at any point x may be calculated from references [16–18].

$$\frac{\delta}{x} = 4.64 \, Re_x^{-0.5} \quad \text{(for laminar flow)} \tag{2.10}$$

or

$$\frac{\delta}{x} = 0.376 \, Re_x^{-0.2} \quad \text{(for turbulent flow)}. \tag{2.11}$$

The local drag coefficient c_f, which enables the local drag force to be calculated, is given by

$$c_{fx} = 1.328 \, Re_x^{-0.5} \quad \text{(for laminar flow)} \tag{2.12}$$

and

$$c_{fx} = 0.0576 \, Re_x^{-0.2} \quad \text{(for turbulent flow)}. \tag{2.13}$$

The overall drag may then be estimated from

$$c_f = \frac{1}{L} \int_0^L c_{fx} \, dx. \tag{2.14}$$

2.2.4.2 INSIDE TUBES AND DUCTS

The boundary layer has zero thickness at the entrance to a tube but builds up through laminar and turbulent modes until the retarded annular region meets at the axis and the flow is said to become "fully developed" (Fig. 2.2). The velocity profile for fully developed laminar flow is parabolic but, once again, the turbulent fully developed velocity distribution undergoes its major change within the laminar sublayer.

The significant dimension for use in the Reynolds number is the hydraulic diameter D_{hd} calculated from

$$D_{hd} = 4 \, \frac{\text{(flow cross-sectional area)}}{\text{(the ``wetted'' perimeter of the pipe)}},$$

i.e. for a tube, $D_{hd} = D$; for an annulus, $D_{hd} = D_2 - D_1$, or for a square duct of side L, $D_{hd} = L$.

The Reynolds number is then $u D_{hd}/v$.

For long ducts, where entrance effects are unimportant, the flow is laminar when $Re < 2100$, transitional when $2100 > Re > 10^5$, and turbulent when $Re > 10^5$.

The pressure drop through a pipe may be calculated from the equation

$$\Delta p = f \frac{L}{D_{hd}} \, \rho \frac{u^2}{2}. \tag{2.15}$$

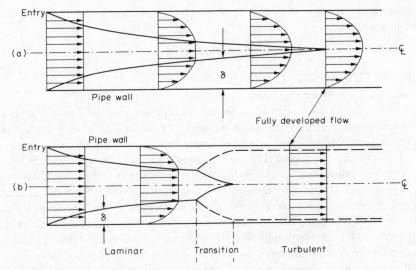

Fig. 2.2. Boundary layer growth at the entrance of a pipe: (a) with a laminar fully developed flow profile; (b) with a turbulent fully developed flow profile.

For fully developed laminar flow the friction factor f is given by the Fanning equation:

$$f = \frac{64}{Re} \qquad (2.16)$$

and is independent of the surface roughness of the pipe (Fig. 2.3). If, however, the heights of surface irregularities are of the order of the thickness of the laminar sublayer, formed in fully developed turbulent flow, the turbulent friction factor is affected. The characteristic relating the friction factor to the flow Reynolds number for turbulent flow inside smooth pipes is [17]

$$f = 0.184 \, Re^{-0.2}. \qquad (2.17)$$

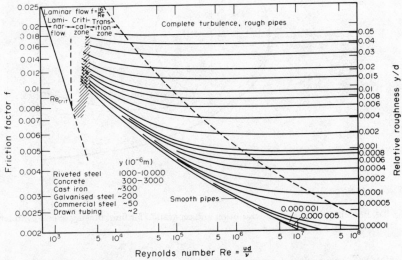

Fig. 2.3. Friction factors *vs.* Reynolds numbers for fluid flowing inside pipes.
(Reproduced with the permission of the American Society of Mechanical Engineers, L. S. Moody, *ASME* Trans, **66**, 1944).

By equating pressure drop and viscous drag offered by the surface, it may be shown that, for an enclosed flowing fluid,

$$f = 4c_f. \tag{2.18}$$

2.2.4.3   OVER TUBES AND CYLINDERS

Boundary-layer growth and separation [17]

When a fluid flows over a solid body it accelerates as it passes over the forward portion and decelerates after passing over the thickest section. The surface boundary layer has zero thickness at the stagnation point on the leading edge. An expanding laminar boundary layer is established over the frontal region (Fig. 2.4) followed by transition to turbulence (at $\sim 80°$ from the stagnation point for a cylinder depending upon the free flow Reynolds number (Fig. 2.5)). The turbulent boundary increases in extent as the distance from the leading edge increases until the positive pressure gradient decreases to zero as the flow accelerates. The boundary layer then separates from the surface at what is known as the "rear stagnation point" [18] (at $\sim 100°$). The region behind the body contains disturbed flow, characterised by eddies, termed the "turbulent wake" or "vortex street". Large pressure losses are associated with flow separation, much energy being dissipated in the wake. For flow over a streamlined body, the pressure drop is caused by skin friction drag. For a bluff body, the skin friction is small compared with the "form drag", arising from separation, which prevents the closing of the streamlines and induces a low-pressure region at the rear of the body. The magnitude of the form drag decreases as the separation point moves further to the rear.

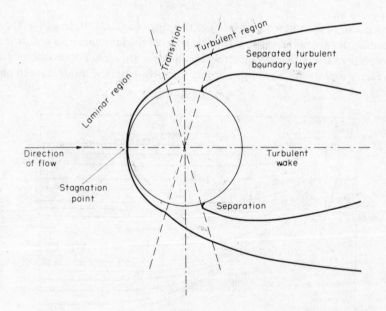

Fig. 2.4. Boundary-layer growth; transition and separation for fluid flowing over a cylinder.

Reynolds number and drag for flow over a cylinder

The Reynolds number is expressed as $u D_o/v$, where D_o is the outside diameter of the cylinder and u is the mainstream velocity. Characteristic flow patterns depend upon the flow Reynolds

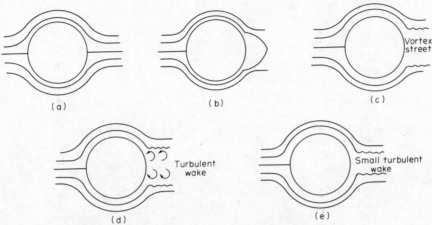

Fig. 2.5. Fow patterns around a cylinder for various flow Reynolds numbers: (a) $Re_D < 1.0$; (b) $Re_D \approx 1.0$; (c) $Re_D \approx 100$; (d) $10^3 < Re_D < 10^5$; (e) $Re_D > 10^5$.

number (Fig. 2.5). The total overall drag coefficient c_d is the sum of the frictional and drag forces and is defined by

$$\text{drag force/unit length} = c_d \frac{\rho u^2}{2} D_o. \tag{2.19}$$

2.2.4.4   TUBE BUNDLES IN CROSS-FLOW

This situation is most commonly encountered in heat exchanger matrices. Because the velocity of the fluid varies as the flow passes over the tubes, a velocity must be selected for use in the Reynolds number. The value in standard use is based on the minimum free flow area available for fluid flow. For an in-line tube arrangement this may be estimated from

$$A_{min} = (s_T - D_o) \times \text{tube length } L, \tag{2.20}$$

whereas, for staggered arrangements,

$$A_{min} = (s_T + D_o/2)L \tag{2.21}$$

or

$$= (s_T^2 + s_{LL}^2)^{\frac{1}{2}} L, \tag{2.22}$$

whichever is the smaller. s_T and s_{LL} are the transverse and longitudinal pitches (with respect to flow direction) of the tube matrix (Fig. 2.6). The pressure drop for an enclosed tubebank can be roughly estimated from eqn. (2.15), although empirically based correction factors are available from manufacturers.

2.2.5   **Energy equations**

Energy may be contained within a flowing system as potential head, velocity, or "kinetic" head $u^2/2g$, pressure head $p/\rho g$, or as heat $c_p T/g$. The total head inside the system boundary is constant at all parts of the system provided no energy crosses the boundary (i.e. an "adiabatic" system). Energy may, however, be interchanged among the different modes. In a nozzle, pressure head is converted to velocity head. At the stagnation point in an immersed body, velocity head is converted to pressure head. The energy lost via pipeline pressure drop, due

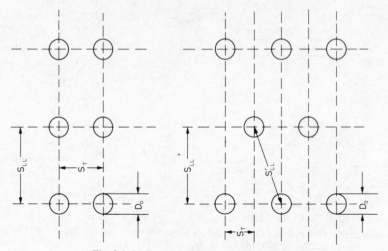

Fig. 2.6. In-line and staggered tube arrangements.

to viscous drag, is in most cases converted to heat as work is performed in overcoming shear forces. The energy loss can also be manifested as noise which itself finally ends up as heat.

$$\text{Total head} = \text{height} + \frac{u^2}{2g} + \frac{p}{\rho g} + c_p T/g \quad \text{(m)} \tag{2.23}$$

or

$$\text{Total energy } E = \dot{m} \left(g \text{ height} + \frac{u^2}{2} + \frac{p}{\rho} + c_p T \right) \quad \text{(W)}. \tag{2.23a}$$

Various guidebooks [19, 20] list pressure losses, total head losses, correction factors (to apply to pressure drops calculated from the fanning equation), of equivalent lengths of ductwork fittings to facilitate the rapid estimation of the energy requirements of complex ductwork systems.

2.3   CONVECTIVE HEAT TRANSFER

Convection is a process of energy transport by the combined action of heat conduction, energy storage, and mixing motion in a fluid. Rates of convective heat transfer depend upon the speed and nature of the fluid flow. If the external force necessary to create fluid movement is caused by a pump or blower, *forced convection* ensues, whereas buoyancy forces, arising from temperature—and hence density—differences, cause *free* or *natural convection*.

2.3.1   The convective heat transfer coefficient h_c

The rate of heat transfer between a surface and a fluid may be computed from

$$\dot{Q} = h_c A\Delta T, \tag{2.24}$$

where the value of the mean convective heat transfer coefficient h_c over the surface under consideration depends upon the geometry of the system, the velocities and modes of fluid flow present, the transport properties of the fluid, and sometimes (i.e. in free convection) upon the magnitude of the temperature difference ΔT. Local values of the heat transfer coefficient h_c are therefore not constant or uniform over the surface.

Whereas local heat transfer rates can be estimated from

$$d\dot{Q} = h_c \, dA \, (T_{sf} - T_\infty), \tag{2.24a}$$

the average value $\bar{h}_c$ is obtained by summing local values over the surface, i.e.

$$\bar{h}_c = \frac{1}{A} \int \int_A h_c \, dA. \tag{2.25}$$

Orders of magnitude of local coefficients

	h_c (W m^{-2} K^{-1})
Air in free convection	5–25
Air in forced convection	25–250
Oil in forced convection	50–150
Water in forced convection	250–10 000

Thermal conductance C is defined as

$$C = hA. \tag{2.26}$$

Thermal resistance of unit area R is the reciprocal of C, i.e.

$$R = \frac{1}{hA}. \tag{2.27}$$

Any analysis which leads to a value for h must start from a study of the dynamics of fluid flow. When fluid is in contact with a surface at a different temperature from that of the bulk of the fluid, heat initially flows from the wall to adjacent fluid particles by molecular conduction. The heated particles are then transported by the action of the flow until a colder region is reached and heat is transferred by conduction to other particles.

In laminar flow, heat is exchanged between adjacent layers by molecular conduction on a submicroscopic scale. In turbulent flow, the conduction mechanism is modified and aided by the eddies which carry lumps of fluid (and therefore their heat contents) across the streamlines. Because of this, turbulent heat transfer coefficients are far greater than those encountered in laminar flow. The parameter ε_h, the eddy diffusivity, is analogous to the eddy viscosity ε_m, being defined by

$$\frac{Q}{A} = -\rho \, c_p \, (\alpha + \varepsilon_h) \, \frac{dT}{dy}. \tag{2.29}$$

The region in the vicinity of a heated or cooled surface where the temperatures differ by less than 99% of the temperature of the free stream T_∞, is called the *thermal boundary layer* of thickness δ_{th} at any point (see Fig. 2.1); e.g. for forced convective laminar over a flat plate,

$$\left(\frac{\delta_{th}}{\delta}\right)^3 = \frac{1}{1.075 \, Pr} \tag{2.29}$$

whereas for turbulent flow over a flat plate,

$$\delta \simeq \delta_{th}. \tag{2.30}$$

2.3.2 **The Nusselt number, thermal diffusivity, and the Prandtl number**

The value of the dimensionless parameter termed the "Nusselt number" is indicative of the ratio of the temperature gradient at the wall to the overall temperature drop between the wall and the edge of the boundary layer. The Nusselt number thus indicates conductive and convective heat transport,

$$\overline{Nu} = \frac{\overline{h}_c L}{k_f},$$ (2.31)

where k_f is the fluid conductivity.

The thermal diffusivity of a fluid is a combination of transport properties defined by

$$\alpha = \frac{k_f}{\rho c_p}$$ (2.32)

where c_p is the fluid specific heat at constant pressure.

The Prandtl number Pr is the ratio of the two derived molecular transport properties: the kinematic viscosity ν and the thermal diffusivity α:

$$Pr = \frac{\nu}{\alpha} = \frac{\mu c_p}{k}.$$ (2.33)

The kinematic viscosity determines the rate of momentum transport through the fluid, whilst the thermal diffusivity dictates the rates of heat transmission. Thus the value of the Prandtl number indicates the relative transfer rates of momentum and heat. Velocity profiles and hydrodynamic boundary-layer thicknesses in laminar flow are affected by ν, whereas temperature profiles and the thicknesses of thermal boundary layers depend upon the value of α. If $\nu = \alpha$, $Pr = \nu/\alpha = 1$, temperature and velocity profiles across a boundary layer are similar, and heat and momentum, or shear stress, processes are analogous. For all gases, $Pr \simeq 1$ (see Appendix IV for air). Molten metals have in general low Prandtl numbers, resulting in the transport of heat being much more rapid than the transfer of momentum. Light oils have very large Prandtl numbers and so are more effective in transmitting viscous shear than heat.

A "turbulent" Prandtl number can be defined as

$$Pr_{tb} = \frac{\nu + \varepsilon_m}{\alpha + \varepsilon_h}.$$ (2.34)

Because $\varepsilon_m \gg \nu$ and $\varepsilon_h \gg \alpha$,

$$Pr_{tb} \simeq \frac{\varepsilon_m}{\varepsilon_h},$$ (2.35)

and since an individual turbulent eddy carries both momentum and heat simultaneously, $Pr_{tb} \simeq 1$ for most flowing fluids. In turbulent flow, therefore, the transport rates of heat and momentum are almost identical.

It can be demonstrated that for flow over a flat plate or inside a pipe, because of the analogy between heat and momentum transfer [18],

$$St = \frac{Nu}{RePr} = c_f/2$$ (2.36)

when $Pr = Pr_{tb} = 1.0$.

2.3.2.1 FORCED CONVECTIVE FLUID FLOW OVER A FLAT PLATE

The local value of the heat transfer coefficient is inversely proportional to the boundary-layer thickness and is thus highest at the leading edge of the plate (Fig. 2.7). It then decreases as the laminar boundary layer thickens, until a sudden increase is encountered as transition to turbulent flow takes place further along the surface. Thereafter the heat transfer coefficient decreases as the turbulent boundary layer grows.

In general,

$$Nu \propto Re^n Pr^m. \tag{2.37}$$

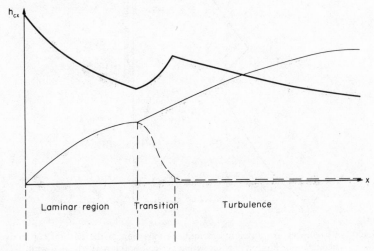

Fig. 2.7. Variation of the magnitude of the surface heat transfer coefficient for fluid flowing in forced convection over a flat surface.

2.3.2.2 FREE CONVECTIVE FLUID FLOW OVER A FLAT PLATE

When an initially stationary fluid is in contact with a hotter solid surface, fluid particles near to the wall are heated by molecular conduction. Buoyancy forces, resulting from temperature-induced density changes, then cause the fluid to move vertically. At steady state, a natural convective boundary layer is formed with viscous forces opposing buoyancy forces (Fig. 2.8).

The fluid in immediate contact with the wall is stationary, being arrested by viscous action. However, as for the forced convective boundary layer, fluid layers further from the surface move progressively faster. A maximum in the velocity distribution is formed between the stationary bulk fluid and the wall, whilst the temperature of the fluid, with respect to that of the surface, increases continuously until it reaches the free fluid temperature. The regions where the temperature and velocity of the fluid are affected by the presence of the wall are known as the hydrodynamic and thermal natural convective boundary layers respectively. Their extents are defined as for forced convective boundary phenomena.

The Grashof number. The Reynolds number has no meaning in natural convection. An alternative dimensionless group, the Grashof number, takes its place:

$$Gr = \frac{\rho^2 g\beta\Delta T L^3}{\mu^2}. \tag{2.38}$$

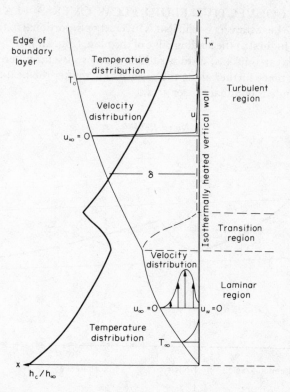

Fig. 2.8. Velocity and temperature distributions and the variation of the natural convective heat transfer coefficient for fluid in contact with a vertical isothermally heated wall.

This represents the ratio of buoyancy forces to viscous forces present. When $Gr \geq 10^9$, transition to turbulence occurs. When considering the walls of rooms in buildings, this change occurs at ~ 1 m from the top or base of the wall depending upon whether ΔT is negative, causing the fluid to fall (downstreaming), or positive, causing the fluid to rise.

Convective heat losses from building surfaces occur mainly by natural convection, although forced convection may become significant with appreciable wind speed. If the ratio of $Gr/Re^2 > 1$, then it is safe to assume that natural convection dominates. Conversely, if $Gr/Re^2 < 1$, a study of forced convective movements and resulting heat transfers should be made. As is demonstrated later, building design for energy conservation should assume an infinite outside surface convection coefficient so that the building can be insulated to cope with the worst wind conditions it may encounter.

In general, for free convective flow,

$$Nu \propto (Gr\,Pr)^n. \tag{2.39}$$

2.3.3   Forced convection inside tubes and ducts

The rate of heat transfer between the inside wall of a pipe and the fluid it contains may be calculated from

$$\dot{Q} = h_c A (T_{sf} - T_b). \tag{2.40}$$

The Nusselt number for a tube is calculated from

$$Nu_D = \frac{\bar{h}_c D_{hd}}{k_f}. \tag{2.41}$$

Temperature and velocity profiles across any flow section in a pipe are qualitatively similar. When $Pr = 1$, these profiles coincide. When $Pr > 1$, $v > \alpha$, $\delta > \delta_{th}$, and the development of the fully developed velocity distribution is nearer the entrance than that for the temperature distribution. The converse applies when $Pr < 1$.

The local heat transfer coefficient depends upon the boundary-layer thickness (Fig. 2.9), being largest at the entrance and decreasing along the pipe until both temperature and velocity profiles are fully established. In laminar flow, entrance effects are significant for distance up to 50 D, whilst fully developed turbulent flow is established at about 10 D from the entrance.

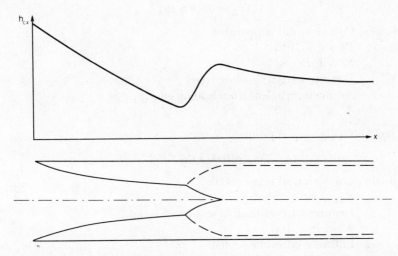

Fig. 2.9. Variation of the local convective heat transfer coefficient at the entrance section of a pipe.

2.3.4 Summary of pertinent Nusselt number relationships

FORCED CONVECTION   LAMINAR FLOW
  Flat plate

$$Nu_x = 0.331 \, Pr^{0.33} \, Re_x{}^{0.5}, \tag{2.42}$$

$$\overline{Nu}_L = 0.662 \, Pr^{0.33} \, Re_L{}^{0.5} \tag{2.43}$$

Conditions: Constant wall temperature.
$\qquad\qquad Re < 3 \times 10^5$.
$\qquad\qquad Pr \sim 1$.
$\qquad\qquad$ Heating starts at leading edge.
$\qquad\qquad T_f = 0.58(T_w - T_\infty) + T_\infty$.
  Inside a pipe. Developing profile:

$$Nu_x = 1.077 (RePr)^{0.33} (D/x)^{0.33}. \tag{2.44}$$

Conditions: Constant wall temperature.

$100 < RePrD/x < 5000$.

Heating starts at entrance.

Fully-developed profile:

$$Nu_D = 1.86 \, (Re_D PrD/L)^{0.33}(\mu_b/\mu_w)^{0.14}. \tag{2.45}$$

Conditions: Constant wall temperature.

$Re_D < 2100$.

Heating starts at entrance.

This equation applies for liquids only.

$T_f = T_b$.

Flat plate

$$Nu_x = 0.0292 \, Re_x^{0.8}. \tag{2.46}$$

$$\overline{Nu}_L = 0.0366 \, Re_L^{0.8}. \tag{2.47}$$

Conditions: Constant wall temperature.

$Re > 5 \times 10^5$.†

$Pr \sim 1, Pr_{tb} \sim 1$.

Heating starts at leading edge.

Assumed turbulent from leading edge.†

$T_f = (T_w + T_\infty)/2$.

Inside a pipe: Fully developed profile:

$$Nu_D = 0.023 \, Re_D^{0.8} Pr^{0.33}. \tag{2.49}$$

Conditions: Constant wall temperature.

$Re_D > 10^4$.

Heating starts at leading section.

$0.5 < Pr < 100$.

Entrance effects neglected.

$T_f = (T_w + T_b)/2$.

Entrance effects. For gases and liquids flowing in short circular tubes $(2 < L/D < 60)$ the heat transfer coefficients obtained for fully developed turbulent flow should be modified according to:

$$\frac{\bar{h}_{cL}}{h_c} = 1 + (D/L)^{0.7} \quad \text{for } 2 < L/D < 20 \tag{2.50}$$

or

$$\frac{\bar{h}_{cL}}{h_c} = 1 + 6 \, D/L \quad \text{for } L/D > 20. \tag{2.51}$$

FREE CONVECTION   LAMINAR FLOW

Vertical flat plate‡

$$Nu_x = 0.508(0.952 + Pr)^{-0.25} Pr^{0.5} Gr_x^{0.25}. \tag{2.52}$$

†A correction for the presence of a laminar section at the leading portion yields

$$\overline{Nu}_L = 0.036 Pr^{0.33} (Re_L^{0.8} - 23\,200) \tag{2.48}$$

‡If the plate is inclined at an angle ψ to the vertical, simply multiply Gr by $\cos \psi$. The same equations may be used.

$$\overline{Nu}_L = 0.677(0.952 + Pr)^{-0.25}Pr^{0.5}Gr_L^{0.25}. \tag{2.53}$$

Conditions: Constant wall temperature.

$GrPr < 10^9$.

$0.01 < Pr < 1000$.

Heating starts at base.

$T_f = (T_w + T_\infty)/2$.

or, alternatively,

$$Nu_x = 0.411(GrPr)^{0.15}. \tag{2.52a}$$

$$Nu_L = 0.548(GrPr)^{0.15}. \tag{2.52b}$$

Horizontal plate

Heated plate facing upwards or cooled plate facing downwards.

$$\overline{Nu}_L = 0.54(GrPr)^{0.25}. \tag{2.54}$$

$10^5 < GrPr < 2 \times 10^7$.

L is the dimension of a side.

$3 \times 10^5 < Gr < 3 \times 10^{10}$.

Heated plate facing downwards or cooled plate facing upwards.

$$\overline{Nu}_L = 0.27 (GrPr)^{0.25}. \tag{2.55}$$

$10^5 < GrPr < 2 \times 10^7$.

$3 \times 10^5 < Gr < 3 \times 10^{10}$.

L is the dimension of a side.

Vertical enclosed airspaces

$$\overline{Nu}_\delta = 0.2 (GrPr)^{0.25}/(L/\delta)^{0.11}. \tag{2.56}$$

$2 = 10^4 < Gr < 2.1 \times 10^3.†$

δ is clearance.

Horizontal enclosed airspaces

$$\overline{Nu}_\delta = 0.21(GrPr)^{0.15}. \tag{2.57}$$

$10^4 < Gr < 3.2 \times 10^5$.

Horizontal cylinders

$$\overline{Nu}_D = 0.53(GrPr)^{0.25}. \tag{2.58}$$

$10^3 < GrPr < 10^9$.

FREE CONVECTION   TURBULENT FLOW

Vertical flat plate

$$Nu_x = 0.0295Gr_x^{0.4}Pr^{0.466}(1 + 0.494Pr^{0.66})^{-0.4}. \tag{2.59}$$

$$\overline{Nu}_L = 0.0246Gr_L^{0.4}Pr^{0.466}(1 + 0.494Pr^{0.66})^{-0.4}. \tag{2.60}$$

†Below $Gr = 2 \times 10^3$, natural convection is suppressed and conduction through the fluid controls, then $h_c = k_f/\delta$.

Conditions: Constant wall temperature.

$GrPr > 10^9$.

Heating starts at base.

Equations apply for turbulent section only.

$T_f = (T_w + T_\infty)/2$.

Horizontal plate

Heated plate facing upwards or cooled plate facing downwards.

$$\overline{Nu_L} = 0.14(GrPr)^{0.33}. \tag{2.61}$$

$$Gr > 3 \times 10^{10}.$$

N.B. Turbulent free convective flow does not occur for the heated plate facing downwards or the cooled plate facing upwards.

Vertical enclosed air spaces

$$\overline{Nu_\delta} = 0.071(GrPr)^{0.33}/(L/\delta)^{0.11}. \tag{2.62}$$

$$2.1 \times 10^5 < Gr < 1.1 \times 10^7.$$

Horizontal enclosed air spaces

$$\overline{Nu_\delta} = 0.075(GrPr)^{0.33}. \tag{2.63}$$

$$3.2 \times 10^5 < Gr < 10^7.$$

2.3.5 Forced convection in flow over tubes

The local conductance is largest at the stagnation point and decreases with distance around the surface as the laminar boundary layer becomes thicker (Fig. 2.10). The heat transfer coefficient reaches a minimum at the sides of the cylinder near the transition region. After an

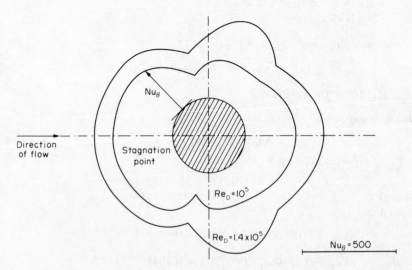

Fig. 2.10. Approximate variation of the local convective Nusselt number for fluid flowing over a tube. (Values of Nu_θ are measured radially from the surface of the cylinder. Angles are referred to the stagnation point at the leading edge.)

increase, as the boundary layer turns turbulent, a further decrease to the point of separation ensues. Over the rear of the body the unit conductance reaches another maximum in the wake.

The variation of the heat transfer coefficient over the surface is thus complex. However, in many instances it is not necessary to know the local values, it being sufficient to evaluate the average value of conductance around the body (Fig. 2.11). The data *for air* can be correlated by

$$\overline{Nu} = K_1 (Re)^n, \tag{2.64}$$

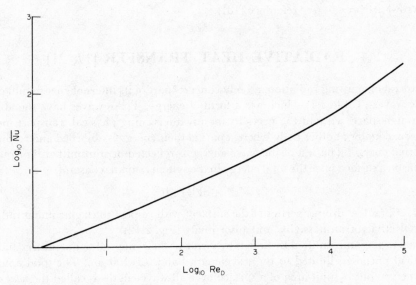

Fig. 2.11. Average values of surface Nusselt numbers for transverse fluid flow over cylinders.

where the constant K_1 and index n depend upon the value of the Reynolds number:

Re_D	K_1	n
0.4–4	0.891	0.330
4–40	0.821	0.385
40–4000	0.615	0.466
4000–40 000	0.174	0.618
40 000–400 000	0.0239	0.805

The following expression is suitable for liquids:

$$\overline{Nu} = (0.35 + 0.56Re^{0.5})Pr^{0.31}. \tag{2.65}$$

2.3.5.1   TUBE BUNDLES IN CROSS-FLOW
Laminar flow To account for tube arrangements, correlating equations are written in the form

$$\overline{Nu} = 0.33K \left(\frac{G_{max}D}{\mu} \right)^n Pr^{0.33} \tag{2.66}$$

for $Re < 200$, where Nu is the average Nusselt number for a bank of 10 or more transverse rows, G_{max} is the mass flow rate per unit minimum free flow area, and K and n are empirical constants depending upon the arrangement.

Turbulent flow
  For $Re > 6000$,

$$Nu = 0.33K \left(\frac{G_{\mathrm{max}}D}{\mu}\right)^{0.6} Pr^{0.33} \tag{2.67}$$

for staggered or in-line rows and for 10 or more transverse rows. In many cases $K \simeq 1.0$.

Corrections for entrance effects must be applied to the heat transfer coefficient when less than ten rows are present (see reference [18]).

2.4 **RADIATIVE HEAT TRANSFER** [17, 21]

In the process of emitting radiation, a body converts part of its internal energy into electromagnetic waves (Table 2.1) which are a form of energy. These waves have the ability to move through space, without the necessity for any intervening physical transport medium, until they encounter another body where a part of their energy is absorbed and reconverted into internal energy. The rest of the wave energy is reflected or transmitted. For radiation impinging on a surface since the total energy in the system remains constant,

$$\mathscr{R} + \mathscr{T} + \mathscr{A} = 1.0, \tag{2.68}$$

where $\mathscr{R}$, $\mathscr{T}$, and $\mathscr{A}$ are properties of the surface (with respect to the incoming radiation) termed reflectivity, transmissivity, and absorptivity (Fig. 2.12).

Thermal radiation is emitted by bodies by virtue of their internal temperatures and is, for all practical purposes, limited to the wavelength band $0.1-100$ μm. The total amount of radiation given out by unit area of a surface over all wavelengths is called the *total emissive power* $\dot{E}_{tot}$. The magnitude of $\dot{E}_{tot}$ depends upon the temperature of the surface. The quality

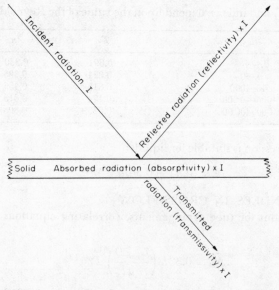

Fig. 2.12. Specular radiation encountering a solid medium.

TABLE 2.1. ELECTROMAGNETIC WAVE SPECTRUM

Wavelength, λ (m)	Type of wave	Band				Frequency (Hz $\equiv$ c s^{-1})
10^{-17}						10^{25}
10^{-16}						10^{24}
10^{-15}	Cosmic rays					10^{23}
10^{-14}						10^{22}
10^{-13}						10^{21}
10^{-12}						10^{20}
10^{-11}	Gamma rays					10^{19}
10^{-10} 1 Ångstrom (Å)						10^{18}
10^{-9}	X-Rays					10^{17}
10^{-8}						10^{16}
10^{-7}	Ultraviolet radiation					10^{15}
10^{-6} 1 micron (μm)	Visible light					10^{14}
10^{-5}						10^{13}
10^{-4}	Infrared (heat) radiation			Hertzian Waves	1 Fresnel	10^{12}
10^{-3} 1 millimetre (mm)						10^{11}
10^{-2}		EHF	Radio Frequencies			10^{10}
10^{-1}		SHF				10^{9}
1 1 metre (m)		UHF				10^{8}
10		VHF				10^{7}
10^{2}		HF			1 megacycle	10^{6}
10^{3} 1 kilometre (km)		MF				10^{5}
10^{4}		LF				10^{4}
10^{5}		VLF				10^{3}
10^{6}	Ultrasonic waves			Slow Oscillations		10^{2}
10^{7}						10
10^{8}						1

of radiation from different sources may be characterised in terms of the wavelength at which maximum emission occurs. For instance, the sun, whose effective surface temperature is approximately 6000 K, emits most of its radiation as short waves between 0.1 μm and 0.3 μm, whilst a body at 1500°C produces most of its radiation within the range 1–20 μm. Glass permits solar radiation to pass through almost unimpeded but is opaque to thermal radiation in the long wavelength range radiated by the interior surfaces of buildings (viz. the "greenhouse" effect).

The reflection of radiation may be *regular* (with the angle of reflectance equal to the angle of incidence) or *diffuse* (i.e. when the same spatial energy distribution is observed from any viewing angle [22]. Reflection from a mirror is an example of the former, whilst good quality white paper is a diffuse reflector.

2.4.1 Blackbody radiation

A "blackbody" is defined as an ideal absorber which absorbs all radiation presented to it (i.e. $\mathcal{R} = 0$, $\mathcal{T} = 0$, $\mathcal{A} = 1.0$) and which also emits the maximum possible amount of thermal radiation at all wavelengths according to its temperature. It is used as a standard with which the radiant characteristics of other bodies may be compared. The emissivity ε of a surface makes this comparison.

The total energy radiated per unit surface area over all wavelengths is given by

$$\dot{E}_{r,bl,tot} = \sigma T^4, \tag{2.69}$$

where σ is the Stefan–Boltzmann constant ($= 5.67 \times 10^{-8}$ W m^{-2} K^{-4}).

The distribution of this radiant energy over the spectrum (i.e. the rate of emission at each wavelength) is given by Planck's law:

$$\dot{E}_{r,bl,\lambda} = \frac{K_1 \lambda^{-5}}{e^{K_2/\lambda T} - 1}, \tag{2.70}$$

where $K_1 = 3.74 \times 10^8$ W m^{-2} μm^{-4}, $K_2 = 1.44 \times 10^4$ μm K, and $\dot{E}_{r,bl,\lambda}$ is expressed in W m^{-2} μm^{-1} with T in degrees Kelvin. The monochromatic (i.e. at a single wavelength) emissive power for various temperature blackbody radiators is plotted in Fig. 2.13 as a function of wavelength. It may be seen that the wavelength at which maximum emission occurs shifts with increasing temperature to shorter wavelengths. The relationship between the wavelength at which $\dot{E}_{r,bl,\lambda}$ is a maximum is described by Wien's displacement law:

$$\lambda_{max} T = 2900 \ \mu\text{m K}. \tag{2.71}$$

Then the maximum monochromatic emissive power at any temperature is given by

$$\dot{E}_{r,bl,max} = \frac{K_1 \lambda_{max}^{-5}}{e^{K_2/\lambda_{max} T} - 1}. \tag{2.72}$$

Equation (2.70) may be expressed alternatively as

$$\frac{\dot{E}_{r,bl,\lambda}}{T^5} = \frac{K_1}{(\lambda T)^5 (e^{K_2/\lambda T} - 1)}. \tag{2.73}$$

This relationship is illustrated in Fig. 2.14. It is seen that peak emission corresponding to

$$\left(\frac{\dot{E}_{r,bl,\lambda}}{T^5} \right)_{max} = 1.28 \times 10^{-11} \text{ W m}^{-2} \ \mu\text{m}^{-1} \text{ K}^{-5}$$

occurs at $T = 2900 \ \mu$m K.

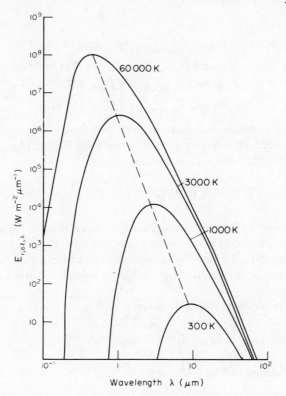

Fig. 2.13. Spectroradiometric curves for an ideal radiator or blackbody at various temperatures: ――― locus of $(E_{bl,\lambda})_{max}$ at $\lambda_{max} = 2900/T$.

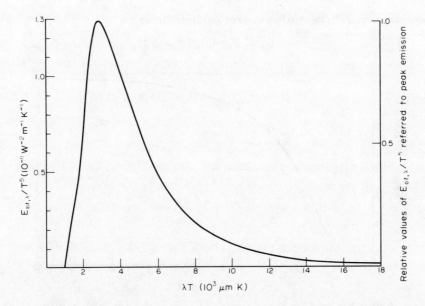

Fig. 2.14. Variation of monochromatic radiant emission with λT and T^5 for various blackbody radiators.

Thus

at   300 K, $\lambda_{max} = 9.66$ μm and $\dot{E}_{r,bl,max} = 31.1$ W m^{-2} μm^{-1}

at 1000 K, $\lambda_{max} = 2.9$ μm and $\dot{E}_{r,bl,max} = 1.28 \times 10^4$ W m^{-2} μm^{-1}

at 3000 K, $\lambda_{max} = 0.966$ μm and $\dot{E}_{r,bl,max} = 3.11 \times 10^6$ W m^{-2} μm^{-1}

at 6000 K, $\lambda_{max} = 0.483$ μm and $\dot{E}_{r,bl,max} = 9.95 \times 10^7$ W m^{-2} μm^{-1}

These figures may also be deduced from Fig. 2.13.

The Stefan–Boltzmann equation may be obtained from Planck's law by integrating eqn. (2.70) over all wavelengths between 0 and ∞, i.e.

$$\int_0^\infty \dot{E}_{r,bl,\lambda} \, d\lambda = \sigma T^4. \tag{2.74}$$

The total area enclosed by any spectroradiometric curve (Fig. 2.13) represents this total energy radiated by a blackbody at the temperature considered.

Many calculations require estimates of energy radiated within a finite band of wavelengths. These may be obtained by integrating eqn. (2.70) between $\lambda = 0$ and λ and dividing by the total emission σT^4, i.e.

fraction of total radiation emitted between wavelengths 0 and $\lambda =$

$$\int_0^r \dot{E}_{r,bl,\lambda} \, d\lambda \bigg/ \int_0^\infty \dot{E}_{r,bl,\lambda} \, d\lambda. \tag{2.75}$$

These fractions are plotted in Fig. 2.15.

2.4.2 Grey surfaces

Radiation from real surfaces is modified by the value of the emissivity, i.e.

$$\dot{E}_{r,grey} = \varepsilon \sigma T^4. \tag{2.76}$$

Heat exchange between two surfaces can be estimated from

$$\dot{Q}_r = \mathscr{F}_{12} A_1 \sigma (T_1^4 - T_2^4), \tag{2.77}$$

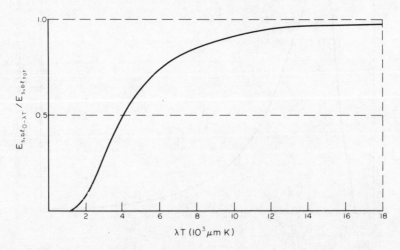

Fig. 2.15. Fractions of total blackbody radiant emissive power between 0 and λT as a function of λT.

where $\mathscr{F}$ is a configuration factor, the value of which (0.0–1.0) depends upon the emissivity of each surface and the relative view (or "shape") factor F_{12} between them (Fig. 2.16). It can be shown that [18] for radiant heat exchange between two "non-black" surfaces at different temperatures,

$$A_1 \mathscr{F}_{12} = \left[\frac{1}{A_1}\left(\frac{1}{\varepsilon_1} - 1 \right) + \frac{1}{A_2}\left(\frac{1}{\varepsilon_2} - 1 \right) + \frac{1}{A_1 F_{12}} \right]^{-1}, \tag{2.78}$$

i.e. for radiation transfer between two parallel flat plates of infinite extent

$$\mathscr{F}_{12} = \left[\frac{1}{\varepsilon_1} + \frac{1}{\varepsilon_2} - 1 \right]^{-1} \tag{2.79}$$

or, for a small, grey body in black surrounding, $\mathscr{F}_{12} = \varepsilon_1$.

2.4.3 Radiative heat transfer with respect to buildings

The evaluation of a configuration factor is often difficult and tedious. Many problems encountered in the analyses of the thermal balances of buildings can, however, often be simplified using average effective radiative heat transfer coefficients $\bar{h}_r$ which may be applied in a similar manner as convective heat transfer coefficients, i.e.

$$\dot{Q}_r = \bar{h}_r A (T_1 - T_2) \equiv A\mathscr{F}\sigma(T_1^4 - T_2^4) \tag{2.80}$$

or

$$\bar{h}_r = \mathscr{F} \left[\frac{\sigma(T_1^4 - T_2^4)}{T_1 - T_2} \right]. \tag{2.81}$$

Now,

$$\left(\frac{T_1^4 - T_2^4}{T_1 - T_2} \right) = (T_1^2 + T_2^2)(T_1 + T_2).$$

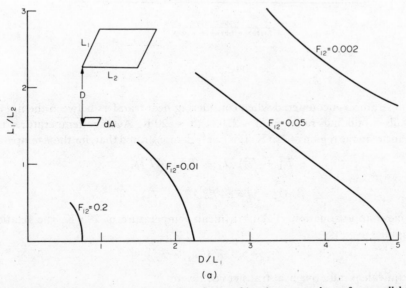

Fig. 2.16. Radiation shape factors: (a) for a surface element dA and a rectangular surface parallel to it ($F_{12} \equiv F_{dA_1 - A_2}$); (b) for adjacent rectangles in perpendicular planes; (c) for equal and parallel rectangles.

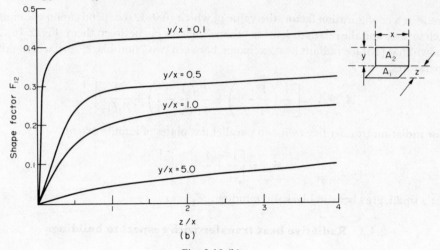

Fig. 2.16 (b).

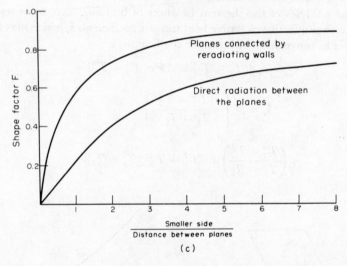

Fig. 2.16 (c).

Surface temperatures encountered when considering heat transfers between the internal or external walls of buildings range from ~270 K to ~320 K. Average temperatures present in the system are in the region of 295 K. It is easily demonstrated that, for these temperatures,

$$(T_1^2 + T_2^2)(T_1 + T_2) \simeq 4T_{AV}^3. \tag{2.82}$$

Thus

$$\dot{Q}_r = A\mathscr{F}\sigma 4T_{AV}^3(T_1 - T_2) \tag{2.83}$$

renders a close approximation. Taking a mean temperature of 295 K, the relationship further simplifies to

$$\dot{Q}_r = 5.8\,\mathscr{F}A(T_1 - T_2) \tag{2.84}$$

giving an equivalent radiative heat transfer coefficient

$$h_r = 5.8\,\mathscr{F} \quad (\text{W m}^{-2}\,\text{K}^{-1}). \tag{2.85}$$

This can then be added to the convective coefficient to produce a heat transfer coefficient which combines the effects of convection and radiation, e.g.:

(I) For two parallel flat plates (i.e. a cavity wall) with inside surface emissivities of 0.9, $\mathscr{F}_{12} = 0.82$ (from eqn. (2.79)), and so $h_r = 4.756$ W m^{-2} K^{-1}, which compares with $h_r \simeq 5.0$ W m^{-2} K^{-1} for the convective coefficient. Thus, in many instances, the convective heat transfer coefficient may simply be doubled to include radiative transfer.

(II) A conservative overestimate of the heat lost from a surface at a temperature T to the sky at night (at an effective temperature of $\sim -45°C$ [17] for a cloudless clear sky) can be obtained using $\mathscr{F}_{12} = \varepsilon$ in eqn. (2.78). For $\varepsilon = 0.9$ and $T = 10°C$, this yields a heat loss per unit area of 191 W m^{-2} for surfaces with a direct view of the sky. This represents an equivalent radiative heat transfer coefficient of ~ 3.5 W m^{-2} K^{-1}. Because the sky temperature is lower than the air temperature, to which convective losses take place (say 0°C), the heat transfer coefficient to be added to the convective coefficient to apply to the surface-to-air temperature difference across the surface boundary layer is proportionately greater (i.e. 19.1 W m^{-1} K^{-1} for an air temperature of 0°C).

It is important to note that, when considering radiative heat transfers, all temperatures must be expressed in degrees absolute. (Further simplifying procedures for the estimation of solar radiation gains to buildings are demonstrated in Chapter 6.)

2.5 CONDUCTION

Conductive heat flow is a process by which thermal energy is transmitted by direct molecular communication. It is the only mechanism by which heat flows in an opaque solid. Conduction in a translucent solid is accompanied by radiation, whilst heat transfer through stagnant gases and liquids takes place by conduction with some radiation. Convection enhances the thermal equilibrium process in moving fluids. The thermal conductivity k of a substance determines its ability to conduct heat.

2.5.1 Steady-state one-dimensional conduction

Because most heat conducted through the walls of buildings flows in a direction perpendicular to the surfaces, most transmission loss or gain calculations involve the one-dimensional Fourier equation for heat flow by conduction:

$$\dot{Q} = kA\frac{\Delta T}{L} = \frac{\Delta T}{R} \tag{2.86}$$

where R (K W^{-1}) is the thermal resistance opposing heat flow by conduction. Its reciprocal C is the thermal conductance of the solid section. ΔT is the temperature drop across a wall of width L and area A.

2.5.2 Composite walls

If n layers of different materials in series comprise a composite wall, the total resistance is calculated by simply summing resistances, i.e.

$$R_{\text{tot}} = \frac{L_1}{k_1 A} + \frac{L_2}{k_2 A} + \frac{L_3}{k_3 A} + \cdots + \frac{L_n}{k_n A} \tag{2.87}$$

and then

$$Q = \frac{\Delta T}{R_{\text{tot}}}. \tag{2.88}$$

If n layers of different materials in parallel make up a wall, the total conductance is calculated from

$$C_{\text{tot}} = \frac{A_1 k_1}{L_1} + \frac{A_2 k_2}{L_2} + \frac{A_3 k_3}{L_3} + \ldots + \frac{A_n k_n}{L_3} \tag{2.89}$$

and then

$$\dot{Q} = C_{\text{tot}} \Delta T. \tag{2.90}$$

2.5.3 **Radial flow conduction**

Because the area A, through which heat passes, increases with radius, a logarithmic mean thermal resistance obtained from integration is employed in radial flow situations. For a pipe of length L

$$R = \frac{\log_e(r_o/r_i)}{2\pi k}. \tag{2.91}$$

Then $\dot{Q} = \Delta T/R$ as before.

Resistances are simply added to obtain the overall resistance of composite pipes.

2.5.4 **Transient heat conduction**

The boundary conditions of most commonly encountered thermal structures are continually changing. For this reason, steady-state analyses provide only approximate solutions. The analysis of heat transmission in the transient mode is complex. In general the heat flowing between a system and its surroundings alters the amount of heat stored within the boundaries. An approximate indication of the time-dependent response may therefore be deduced by making the assumption of negligible internal resistance (i.e. that the temperature of the system is uniform at any instant).

If this temperature changes by an amount dT during a time interval dt, the change in internal energy is equal to the net heat flow rate across the boundary, i.e.

$$c\rho V \, dT = \hbar A (T - T_\infty) dt, \tag{2.92}$$

where c and ρ are the specific heat and density of the material comprising the structure, V is its volume, A the area of the boundary, $\hbar$ the heat transfer coefficient at the boundary, and T_∞ the temperature of the surroundings. By separation of variables and integration, the transient temperature response is obtained as

$$\frac{T_t - T_\infty}{T_{t=0} - T_\infty} = \exp\left[-\left\{\frac{\hbar A}{c\rho V}\right\}t\right], \tag{2.93}$$

where $c\rho V/\hbar A$ is called the "time constant" Θ of the system. Its value is indicative of the time taken for the difference between the system temperature and that of the surroundings to alter to 36.8% (i.e. $e^{-1} \times 100\%$) of the initial temperature difference.

The Biot modulus Bi is a dimensionless group which represents physically the ratio of the internal thermal resistance of the system to the external resistance at the boundary.

$$Bi = \frac{hL}{k_s}, \tag{2.94}$$

where k_s is the thermal conductivity of the solid material and L is obtained by dividing the volume of the solid by its surface area. The error introduced by the assumption that the temperature is uniform at any instant of time is less than 5% when $Bi < 0.1$. The considerations outlined form the basis of approximate procedures available to investigate the transient behaviour of buildings (Chapter 7).

2.6 MASS TRANSFER AND PSYCHROMETRY

Atmospheric air always contains a portion of water vapour (~ 0–0.03 kg/kg of dry air). This is a small fraction of the total mass and, because the transport properties of water vapour are of the same order as those of dry air, little error results in many sensible heat transfer calculations if the presence of water is neglected. If, however, a change of phase in the water constituent occurs (i.e. evaporation or condensation), the concomitant latent heat transfers contribute significantly to the overall exchange of heat. If, for example, 1 kg of saturated air containing 0.03 kg of water vapour were to cool by 10°C, the latent heat rejected by the vapour in condensing (75 kJ) is many times that lost by the dry air (10 kJ) in the process of sensible cooling [23]. Any insulated thermal system must therefore prevent phase changes at its boundaries. The phenomena can, on the other hand, be suitably employed in some enhanced heat transfer applications (i.e. heat pipes or heat pumps (Chapters 10 and 11)). If a phase change occurs inside a boundary layer, the resulting instantaneous heat transfer coefficient can be up to 100 times that for natural convection alone.

2.6.1 Psychrometry

Psychrometry applies the ideal gas laws separately to each of the two substances (water and air) present in an atmospheric mixture to produce tables and charts to aid design calculations.

Dalton's law of partial pressures states that if a mixture of two gases occupies a given volume at a certain temperature, the total pressure exerted by the mixture is equal to the sum of the pressures contributed by each constituent gas. The characteristic gas law states

$$pV = mRT, \tag{2.95}$$

where p (Nm^{-2}) is the pressure exerted by m (kg) of the gas at a temperature T (K) contained within a volume V (m^3) and $\mathbf{R}$ is the characteristic gas constant.

For dry air, $\mathbf{R}_a = 287 \, J \, kg^{-1} \, K^{-1}$, whilst, for water vapour, $\mathbf{R}_v = 462 \, J \, kg^{-1} \, K^{-1}$.

Water vapour may be treated as a perfect gas which is, however, capable of liquefaction by the application of pressure alone at common atmospheric temperatures. Pure liquid water boils at 100°C under an atmospheric pressure of 101 325 N m^{-2} ($= 1013.25$ mbar). If the total pressure is reduced, boiling occurs at lower temperatures (see steam tables). The relationship between pressure and temperature at which the change of phase from liquid to vapour occurs is known as the "saturation curve" (Fig. 2.17).

Liquid water at a state point located on the saturation curve will evaporate at constant temperature and pressure with the application of heat. Thus a psychrometric mixture of water and air at a given temperature can hold no more water vapour than the corresponding saturation pressure will permit, i.e. if

$$T = 30°C, \, p_{sat} = 4240 \, N \, m^{-2},$$

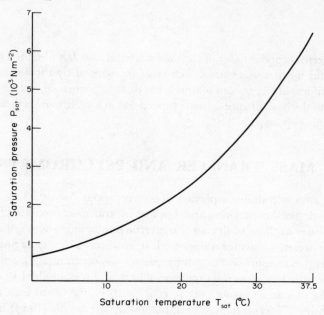

Fig. 2.17. Saturation curve for water.

and, from eqn. (2.89),

$$\gamma = \frac{m_v}{m_a} = \left(\frac{pV}{\mathbf{R}T}\right)_v \left(\frac{\mathbf{R}T}{pV}\right)_a = \frac{p_v \mathbf{R}_a}{\mathbf{R}_v p_a}.$$

$$= \frac{4240}{101\,325 - 4240} \frac{287}{462}$$

$$= 0.0274 \text{ kg moisture per kg dry air.} \tag{2.96}$$

The ratio γ is known as the *specific* humidity, or moisture content. If more than the amount of water required to saturate the air is present, the remainder will not evaporate. If, however, less than this amount is available, all will evaporate providing that enough heat is supplied to fuel the process. The resulting unsaturated mixture will have a *relative humidity* ϕ given by

$$\phi = \frac{p_v}{p_{sat}} = \frac{(m\mathbf{R})_v}{(m\mathbf{R})_{v,sat}} \tag{2.97}$$

in which p_v is the vapour pressure of the liquid.

When a psychrometric mixture is cooled at constant pressure and specific humidity, its relative humidity rises until it reaches 100% at the saturation curve. Further cooling then causes droplets to condense from the air. The temperature at which this occurs is the saturation temperature corresponding to the mixture's specific humidity. For convenience this is termed the "dew-point" temperature, i.e. when

$$\gamma = \frac{m_v}{m_a} = \frac{p_{sat}}{\mathbf{R}_v} \frac{\mathbf{R}_a}{p_a}. \tag{2.98}$$

The psychrometric chart (Fig. 2.18) has dry-bulb temperature, T_{db} and specific humidity as ordinates. The wet-bulb temperature T_{wb} is a value indicated by an ordinary thermometer

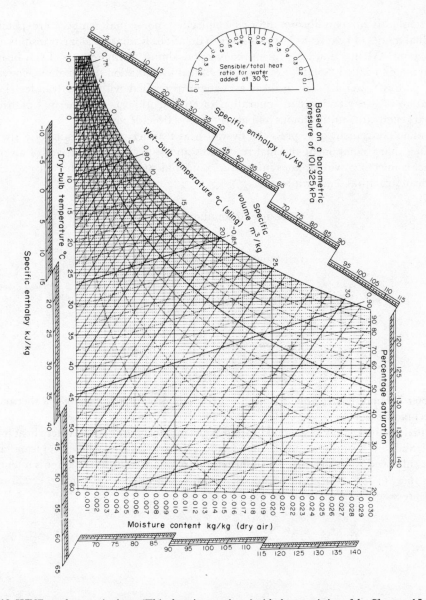

Fig. 2.18. IHVE psychrometric chart. (This chart is reproduced with the permission of the Chartered Institution of Building Services. Pads of charts size A3 suitable for permanent records are obtainable from the Institution.)

with a water-moistened wick wrapped around the bulb. Evaporation of water extracts sensible heat from the bulb, and at steady state causes the device to indicate a temperature lower than the dry-bulb temperature. This easily measured parameter enables the state point of the air to be located on the chart. Values for the relative humidity, dew-point temperature, specific volume, and enthalpy may then be deduced, and the effects of various processes on the relative amounts of liquid water and vapour may be predicted.

2.7 **HEAT TRANSFER BY COMBINED MODES**

Heat transfer is an equilibrium process connecting non-equal temperature potentials. Individual modes of heat transfer seldom act in isolation. Radiation alone transmits energy in a vacuum. When a fluid is present, convection almost always ensues. Even heat flow through some so-called "solids" are sometimes aided by radiation (i.e. in glass), convection (i.e. in porous materials), or by heat exchanges associated with mass transfer (i.e. the evaporation of sweat and vapour migration through clothing). The thermal paths in an insulated system are multitudinous and nearly always time-dependent.

It is often convenient to express the overall heat transfer characteristics of a structure containing many component thermal resistances and capacitances in terms of an overall resistance connecting two extreme temperature potentials and a lumped thermal capacitance at the mean structural temperature, i.e.

$$\text{resistance to conduction} = \frac{L}{kA},$$

$$\text{resistance to convection} = \frac{1}{h_c A},$$

$$\text{resistance to radiation} = \frac{1}{h_r A},$$

$$\textit{total resistance} = \frac{1}{UA},$$

$$\textit{lumped thermal capacitance} = \sum mc.$$

Chapter 7 demonstrates how this may be achieved for complex systems via thermal network techniques.

Thermal partitions may in general be subdivided into insulating walls, where heat transfer must be inhibited (Chapters 6 and 8), and heat exchanger surfaces, where heat transfer rates must be enhanced (Chapter 10). The information summarised in this chapter is intended to provide most of the data required for the optimisation of these extreme thermal energy transfer processes.

Chapter 3

Thermal Comfort

3.1 WHAT IS THERMAL COMFORT?

Thermal comfort is the study of the effects of climatic impact on human response. A state of comfort has been defined [24] as "that condition of mind which expresses satisfaction with the environment". Although a person may sometimes act more efficiently for short periods when subjected to a stimulus of slight discomfort, intellectual, manual, and perceptual performance is in general highest for long periods when the environment can be rated as being comfortable [25]. An ideally thermally comfortable environment is one where the occupants experience no heat stress or thermal strain. The condition is a neutral state in which the body needs to take no particular action to maintain its proper heat balance. In order to optimise on fuel/thermal insulation ratios for the design of heating, ventilating, and air-conditioning systems it is necessary to examine the criteria of human thermal comfort. Unfortunately, thermal comfort has become a commodity produced by the service industries and marketed and sold by the heating, ventilating, air-conditioning, and insulating engineers [26]. Thus, whereas in the past individuals relied upon clothing to maintain thermal equilibrium, recent trends depend upon the production of artificial interior climates. Thermal isolation is therefore purchased more expensively from the building services engineer rather than from the tailor.

3.2 ASSESSMENT OF THERMAL COMFORT

The definition of thermal comfort criteria uses three basic models: the physical, physiological, and sociological approaches.

The physical model considers the body as a thermal system in which heat is exchanged between the body tissues and the environment through the skin and clothing. Equations describing comfort states have been developed [27] for combinations of environmental parameters which result in zero net heat storage within the body.

The physiological model examines subjective responses to imposed thermal environments and produces data for comfortable conditions. Involuntary actions such as sweating, vaso-regulation, and shivering, which occur when the body is outside the neutral state, are also studied. Thus responses may also be predicted for "off-design" environments. The ASHRAE comfort chart (Fig. 3.1) [24] shows combinations of environmental variables for which neutral response has been evoked and also indicates degrees of discomfort produced by surroundings maintained outside the one encompassing these comfortable states.

43

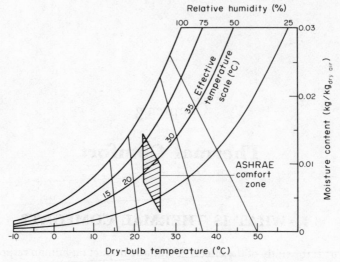

Fig. 3.1. The ASHRAE comfort zone.

The application of accepted human comfort criteria to the environmental conditioning of interiors is often hindered by sociological factors. The flow of metabolic heat to the environment can be voluntarily adjusted by changes in posture or activity, altering clothing insulation levels, or by regulating the thermal environment (i.e. opening a window). The behavioural response of an individual to any particular stimulus depends upon social conditions and constraints [24]. These give rise to pressures which modify the response or place limits on its extent.

An individual human being prefers to have a degree of control over his environment. It follows that the imposition of a rigorously controlled "comfortable" constant thermal condition as indicated by physical or physiological studies may not be acceptable psychologically or socially. Indeed, an allowance for flexibility not only saves energy but can also lead to greater satisfaction.

3.3 OFF-DESIGN CONDITIONS

In a thermally comfortable environment, body heat loss is accomplished primarily by natural convection and radiation. At higher temperatures evaporative loss becomes dominant [29]. When the environmental temperature is above $\sim 33°C$, heat is transferred *to the body* by convection and radiation, and so sweating and evaporation must offset these heat gains as well as that arising from body heat production. Under such circumstances the psychrometric properties of the air become important. If an occupant of a high-temperature environment is so constrained or limited that he can take no subjective action to regulate rates of heat transference to the surroundings, it becomes necessary to produce an externally controlled forced cooling arrangement to avoid physiological strain. Such conditions are encountered in, for example, an aircraft cockpit which can be subjected to up to 1 kW m^{-2} of solar heating, or in enclosed ground vehicles or submarines where internal temperatures can reach over 60°C with high relative humidities unless adequate ventilation and/or air conditioning is provided.

The resulting thermal situations are not strictly "comfortable" as defined by thermal comfort criteria but are intended to avoid appreciable thermal strain. It is not surprising that most physiological studies of extreme stress conditions have been undertaken with respect to military vehicles. However, even in the United Kingdom, mean effective environmental temperatures in offices can reach 40°C just inside a south-facing glazed wall. High heat exposures are tolerated by furnace workers, foundrymen, steelmakers, bakers, and brick-workers who commonly maintain thermal balance by active sweating accompanied by accelerated heart rates. In industrial environmental control problems some heat exposure in excess of comfort conditions must usually be accepted. The worker must endure some heat stress but this should be kept below the hazardous level.

3.4 EFFECTS OF ENVIRONMENTAL VARIABLES

The human body is a heat engine which burns food and oxygen and converts the energy released into work and heat; the harder the work rate the more metabolic heat is generated and the faster the rate of heat transfer to the environment. The body is seldom in a state of steady heat transfer with its surroundings; this only occurs when bodily exercise rates and environmental conditions are relatively stable. The effective environmental temperature and its temporal fluctuations are registered by nerve-endings beneath the skin. The neutral skin temperature is of the order of 33°C and sensations of warmth or cold are produced when the environmental temperature is above or below this figure. The body then takes action to regulate heat transmission voluntarily or involuntarily. The principle factors affecting the comfort sensation are:

air temperature
radiant temperature
air velocity
relative humidity
clothing level
degree of activity

Variation in mean air temperature affect the rate of heat transfer by convection, whilst gains and losses by radiation are controlled by the mean radiant temperature of surrounding surfaces. These "dry" heat exchange modes are assisted by the "latent" heat contribution when sweat exuded from the pores of the skin evaporates to the surroundings and, in the process, absorbs the required latent heat of vaporisation from the mass of the body. Rates of evaporative cooling depend, therefore, upon the difference between the partial pressure of the water vapour at the skin's surface and that prevailing in the surrounding air. Thus the body is able to reject great quantities of heat even when the ambient air and radiative environmental temperatures are above the skin temperature. Evaporative losses occur in two forms:

(a) Passive water loss from the lungs by respiration and diffusion through the skin at normal temperatures. This can contribute up to 25% of the overall dispersion of metabolic production in the comfort state.

(b) At elevated environmental temperatures active sweating by secretion from the sweat glands occurs, dispersing 100% of metabolic heat and retransmitting heat gains from the environment.

Air movement affects both convective and evaporative heat transport systems.

When the surrounding temperature approaches blood heat ($\sim 37°C$), a high humidity becomes dangerous as it prevents the effective operation of the evaporative cooling mechanism.

Whereas in a dry-air environment it is just possible to endure a temperature of $\sim 52°C$ in the shade, if the air is saturated the maximum endurable temperature is only $\sim 33°C$. There are, of course, other factors which can considerably reduce these figures: a clothed human, physically active in direct sunlight, could suffer heat stroke in an environment at only 20°C (db) at 100% relative humidity [30].

The total heat loss from an adult at rest is approximately 100 W at normal room temperatures. This consists of ~ 75 W of sensible heat and ~ 25 W of latent heat. The latent heat loss arises from the evaporation of ~ 0.05 kg h^{-1} of water.

3.5   PARAMETERS AFFECTING THERMAL COMFORT

3.5.1   Rate of metabolism Q_M

This is the body's principle source of heat resulting from the oxidation of food elements within the body. A human being needs a food input of ~ 0.15 kW for basal survival, although greater quantities are required to replace energy expended as work. A net loss in body heat cannot be achieved by mechanical expenditure since this violates the Carnot principle. The basal metabolic rate is about 40 W m^{-2}. Light exercise produces ~ 50 W of work for a metabolic rate of ~ 120 W, resulting in a thermal efficiency of $\sim 40\%$. The figures for heavy work are ~ 200 W, ~ 300 W, and $\sim 66\%$ respectively [24]. Rates of metabolism are easily measured by monitoring rates of oxygen consumption. Table 3.1 lists ranges of values found for various activity levels.

TABLE 3.1. METABOLIC RATES FOR DIFFERENT ACTIVITY LEVELS [24]

Activity	Rate of heat production (W m^{-2})	Activity	Rate of heat production (W m^{-2})
Sleeping	40	Machine work	100–260
Seated quiet	60	Shop assistant	120
Walking (3 mph)	150	Teacher	90
Light work	120	Vehicle driving	80–180
Medium work	170	Domestic work	80–200
Heavy work	300	Office work	60–80
Heaviest work possible	450–500		
		Tennis	200–270
Carpentry	100–370	Squash	290–420
Foundry work	170–400	Wrestling	400–500
Garage work	80–170	Golf	80–150

NB—The average body area is approximately 1.8 m² [24].

3.5.2   Body temperatures

The mean deep body temperature must be maintained constant at 36–38°C for the effective operation of its component organs. The oesophageal temperature is that of the blood in the heart. The bloodstream carries heat to all parts of the body to maintain the necessary uniform temperature. The temperature attained by the tympanic nerve, situated below the eardrum, provides the thermoregulatory system with the feedback information which stimulates response. Although thermometric devices are available to monitor tympanic temperatures, deep body temperatures are more readily estimated by measuring oral or rectal values.

The mean skin surface temperature $\bar{T}_{sk}$ varies between 23° and 36°C for a clothed human body and can range from 8° to 40°C when nude. It is generally accepted that $\sim 33°$C produces a comfort response. Values can be measured using a radiometer or a surface thermoprobe. The mean skin temperature is usually averaged from ten locations.

3.5.3   **Ambient air temperature** T_a

The dry-bulb temperature of the environmental air is the simplest practical index of cold or warmth under ordinary room conditions. Its use as a single indicator of a comfortable thermal environment is limited when sweating occurs or when high-power radiant sources are present. The measurement of air temperature is relatively straightforward provided that the effects of thermal radiation between the sensor and surrounding surfaces are eliminated. This is most easily accomplished by surrounding the detector with an aspirated radiation shield. In general, the larger the surface area of the probe, the more radiant energy is intercepted. Thus a thermocouple bead is more effective than a glass thermometer for measuring air temperature. Without shielding, however, even a very small thermojunction produces a response which includes $\sim 20\%$ due to radiative heat transfer. The effect of aspiration is to speed up the fluid flow rates and thus increase the relative amount of convective heat transfer between the device and the air. In practice, thermometers, thermocouples, or thermistors are used with the precautions outlined.

3.5.4   **Mean radiant temperature of the environment** $\bar{T}_r$

The mean radiant temperature at a point in an enclosure is a function which is dependent upon the temperatures of all surface areas in thermal view of the point. In thermal comfort applications it refers to the shape and surface radiative characteristics of the human body with respect to surrounding surfaces: it is therefore difficult to measure precisely. Mean radiant temperature is defined as the uniform blackbody temperature of an imaginary enclosure with which an occupant will exchange the same heat by radiation alone as in the actual complex environment. A value for $\bar{T}_r$ can be obtained simply by measuring the temperatures and areas of all surfaces (ceilings, walls, floors, radiant sources) and then using the data so obtained in a relationship of the form

$$\bar{T}_r = \frac{\sum A_j T_j}{\sum A_j},\qquad (3.1)$$

where T_j is the temperature of surface j of area A_j. Thus all radiant temperatures are weighted by their effective areas. This, however, gives a constant value of the mean radiant temperature throughout the room, whereas, strictly, because angle factors between emitting and receiving surfaces depend upon their relative locations, $\bar{T}_r$ is also a function of position [21].

Several instruments are available for the direct measurement of mean radiant temperature at a location in an enclosure in relation to the irradiation of a sphere. These will, in many cases, give a reasonable approximation of the effective temperature with which radiative exchanges between the human body and its environment takes place [31–33]. Because of its simplicity, the globe thermometer is often adopted in practice [34]. This consists of a 150 mm diameter black spherical shell into the centre of which a thermal sensor is placed. This sensor takes on the mean temperature of the globe. The larger the sphere the greater the amount of radiant heat intercepted in relation to heat gains by convection. The value indicated by the globe thermometer is influenced by the dry-bulb temperature and air velocity as well as the radiant temperatures. Thus in using the instrument suitable corrections for convective heat transfers have to be applied.

3.5.5 Air velocity

For comfort studies a single value for air speed usually suffices to describe the cooling or warming effects of the air motion. The range of air velocities encountered indoors is about 0.1 to 1.0 m s^{-1}. If the room circulatory air speed is below ~ 0.1 m s^{-1}, only natural convection takes place. The air velocity around a person usually results from a combination of induced free convective flows and forced air movements. Surface boundary layers are generally turbulent. Measurement of the low velocities present is difficult although several instruments are available. These include electrically heated anemometers and transient devices. The Kata thermometer is the most widely used sensor of low fluid flow rates. This consists of an alcohol-in-glass thermometer with a large (~ 3–6 mm) spherical bulb which is placed in the air stream. The time taken for the alcohol to fall between two graduations marked on the stem is recorded. The air speed may then be calculated from the time constant of the system.

Room air circulatory patterns may be observed using airborne tracers: smoke, feathers, radioactive gases, bubbles, or metaldehyde. Estimates of local velocities can be deduced using photographic techniques.

3.5.6 Humidity

The amount of water vapour present in air may be assessed in various ways by measuring any two of the following parameters: dry-bulb temperature, wet-bulb temperature, relative humidity, dew-point temperature, or specific humidity. The most widely used instrument is the wet- and dry-bulb thermometer combination used in a sling, in a louvred box (for meteorological measurements) or in a mechanically aspirated device.

Wet-bulb temperature T_{wb}. This is the temperature at which water, by evaporating into air, can bring the air to saturation adiabatically at the same temperature. The rate of evaporation of a liquid depends upon the degree of saturation of the surrounding air. When a rapid evaporation occurs, a cooling effect is initiated as the more rapidly moving molecules escape. Thus in the wet-bulb thermometer latent heat of vaporisation is absorbed from the bulb causing a temperature lower than the dry-bulb temperature to be registered.

The wet- and dry-bulb system contains two thermometers, one of which is covered with a wick which is kept wet. Heat flows from the environment to the thermometers by convection and radiation. Providing that the adjacent air velocity is high enough (> 3 m s^{-1}), the dry bulb will take on the temperature of the air. A state of equilibrium is reached for the wet bulb in which the sensible heat gain to the water surrounding the bulb exactly balances the latent heat loss from it. The steady-state temperature achieved is termed the wet-bulb temperature of the surrounding moist air. Meteorological louvred boxes rely upon wind speed to provide the necessary air velocity. For this reason the wet-bulb temperature recorded, termed the "screen" value, is considered to be approximately 1°C higher than values indicated by whirling or aspirated thermometers.

Relative humidity ϕ. Many organic materials experience dimensional change with changes in humidity levels. This action has been utilised in a number of simple and effective calibrated humidity recorders, indicators, and controllers. Materials commonly employed include human hair, nylon, animal membrane, wood, and paper, amplification of the response being usually provided by mechanical or electrical ancillaries. Unfortunately, no material has been found which reproduces readings consistently over a period of time, and the responses may be significantly affected by exposure to extremes of humidity. Thus frequent recalibration is required.

Dew-point temperature T_{dp}. The dew-point is the temperature at which dew would begin to form if the air were slowly cooled. It is, therefore, the temperature at which the saturated water vapour pressure is equal to the prevailing partial pressure of the water vapour in the air. Dew-point temperatures can be determined by noting the temperature of a polished metal surface when the first traces of condensation appear. Commercial devices give very accurate estimates of the psychrometric state by incorporating feedback control to miniature heating/cooling systems triggered by photocells viewing a mirrored surface [35].

Specific humidity. The gravimetric hygrometer is the ultimate standard for calibrating humidity-measuring devices. The water vapour in an air sample is absorbed by suitable chemicals and then the adsorbent is very carefully weighed. Phosphorous pentoxide and magnesium perchlorate produce most accurate measurements whilst silica gel or calcium chloride can also be used.

Other hygrometers. Various devices have been developed which utilise substances whose electrical resistance properties depend upon moisture contents absorbed from ambient moist air (i.e. lithium chloride). These must always be maintained in calibration.

The electrolytic hygrometer allows a continuous supply of sample gas to flow through an analyser tube where the moisture is absorbed by a dessicant (usually phosphorous pentoxide) and the water is then electrolysed into hydrogen and oxygen. Consequently a measurable electrolytic current flows, the magnitude of which is indicative of the moisture content of the air.

3.5.7 Level of clothing

Clothing forms a resistance to the transfer of heat and moisture between the skin's surface and the environment. Factors have been incorporated in human thermal comfort heat balance equations adopted by the ASHRAE [24] to include the effects of clothing levels. Insulation provided by clothing assemblies is discussed in Chapter 8.

3.6 MEAN EFFECTIVE TEMPERATURE INDICES

There are obviously an infinite number of combinations of values of the four primary environmental parameters (air temperature, mean radiant temperature, humidity, and air velocity) which induce a feeling of thermal comfort. It is, therefore, not surprising that many attempts have been made to define and measure a single index which combines the effects of these parameters in thermal comfort subjective assessment studies. Many of these indices have no mathematical significance but give relative values in a qualitative manner only.

3.6.1 Indices obtained from subjective environmental ratings

Subjective assessments of an environmental combination of the four primary parameters may be rated on the seven-point Bedford scale [1]:

(1) Much too cool.

(2) Too cool.

(3) Comfortably cool.

(4) Comfortable.

(5) Comfortably warm.

(6) Too warm.

(7) Much too warm.

Subjects are placed in an environmentally controlled chamber and, as primary parameters are adjusted singly, are asked to vote at hourly intervals. The data are statistically averaged and an effective temperature scale is assigned to the range of conditions imposed during the test. The occupant is thus used as a thermometer which responds to rates of heat loss or gain between the body and its environment, the subjective rating representing the response "read-out". If the environmental humidity is maintained at 50%, the air speed below 0.3 m s^{-1}, and the mean radiant temperature equal to the air temperature, then a lightly clothed person would assign approximately 3°C as the difference between successive points on the scale, being thermally neutral at about 25°C.

The oldest and most widely used index formulated on this basis is the "*old*" *effective temperature* T_{eff} devised in the United States in the 1920s to relate air temperature, humidity, and air speed to the comfort sensation. The numerical value of T_{eff} is the temperature of still air at 100% relative humidity which produces the same feeling of comfort or discomfort as the actual environment. Values were obtained using an environmental cabinet containing adjoining rooms, one of which was maintained at 100% relative humidity and with still air. Test personnel were required to walk from one room to the other as the air temperature in the reference room was varied until they experienced no change in their reaction to the two environments. It is now commonly agreed that in devising the scale the test procedure adopted led to an overestimation of the effects of humidity.

The *subjective temperature* T_{sub} [36] is a more recent attempt to quantify the environment in purely subjective terms. It is defined as the temperature of a uniform enclosure with mean radiant temperature equal to the air temperature (with air velocity equal to 0.1 m s^{-1} and with a relative humidity of 50%) which would produce the same feeling of warmth as the actual environment under consideration. It has been shown that for an air velocity of less than 0.1 m s^{-1} in the comfort region,

$$T_{sub} = 0.56\, T_a + 0.44\, \bar{T}_r. \tag{3.2}$$

Thus the relative effects of mean radiant temperature and air temperature on feelings of human thermal comfort may be deduced. This obviously neglects the effects of thermal asymmetries and clothing levels, and inherent differences between different personal assessments.

3.6.2 Physically derived effective temperatures

As long as all the necessary physical information is available, it is a simple matter to combine the effects of two or more primary environmental variables mathematically.

The operative temperature T_{op} is the temperature of a hypothetical uniform isothermal "black" enclosure in which the human body would exchange the same *dry* heat as is exchanged by radiation and convection as in the actual environment. T_{op} is thus the average of the mean radiant and air temperatures weighted by their respective heat transfer coefficients:

$$T_{op} = \frac{(h_r \bar{T}_r = h_c T_a)}{(h_r + h_c)}.$$ (3.3)

It has been shown (Chapter 2) that, for moderate temperatures, radiative and natural convective heat transfer coefficients are approximately of equal value numerically. For a human being completely enclosed in a thermally black enclosure, the radiant configuration factor is unity. Liquid water has an emissivity of ~ 1.0. It may be shown that, since the human body is composed mainly of water with a surface skin which is largely transparent to infrared radiation, the surface emissivity of the human body also has a value approaching 1.0. A radiant heat transfer coefficient for the transfer system may therefore be deduced (following the procedure outlined in Chapter 2) as ~ 5 W m^{-2} K^{-1}. Brundrett [37] has shown that the mean naturally induced convective heat transfer coefficient for sitting or standing personnel may be estimated from

$$h_c = 2.285(T_{sk} - T_a)^{0.25}$$ (3.4)

whilst an average value, $h_c \simeq 5$ W m^{-2} K^{-1}, has also been quoted [38].

The *humid operative temperature* T_{oh} has also been developed [24] as an effective temperature which combines the effects of air temperature, mean radiant temperature, and humidity. T_{oh} is defined as the uniform temperature of an environment at 100% relative humidity with which a human body would exchange the same rate of heat as is transferred in the actual environment by radiation, convection, and evaporation.

$$T_{oh} = \frac{h_r \bar{T}_r + h_c T_a + h_{ev} T_{dp}}{h_r + h_c + h_{ev}}.$$ (3.5)

This parameter is, however, little used because the evaporative heat transfer coefficient depends upon the degree of skin "wettedness" and so is difficult to quantify numerically.

Both T_{op} and T_{oh} have mathematical significances in calculations of body heat balance. Both, however, conceal the effects of air velocity.

Various sophisticated and specialised instruments have been developed in an attempt to measure combined effect indices. These are intended to simulate the thermophysical characteristics of the human body and measure the total heat loss to an environment corresponding to a person in a state of thermal comfort [39, 40].

3.7 THERMAL EXCHANGES BETWEEN MAN AND HIS ENVIRONMENT

The general heat balance equation describing the steady thermal state of a human body may be expressed as

$$\dot{Q}_{st} = \dot{Q}_M \pm \dot{W} \pm \dot{Q}_{ev} \pm \dot{Q}_r \pm \dot{Q}_c \quad (\text{W m}^{-2}).$$ (3.6)

When the body is in a state of thermal comfort the rate of heat storage $\dot{Q}_{st}$ within it is assigned a value of zero by researchers in physical thermal comfort [27]. A positive or negative value means that the body is gaining or losing heat. This situation cannot continue uncontrolled, and so some regulation of surface heat transfer rates is required to regain the thermally neutral condition.

TABLE 3.2. THE "NEW" EFFECTIVE TEMPERATURE INDEX

ET^* (°C)	Temperature	Sensation Comfort	Physiology	Health
43		Limited tolerance	*Zone of inevitable body heating*. Failure of regulation, increasing heat stress	Circulatory collapse
40	Very hot	Very uncomfortable	*Evaporative regulation against heat*. Sweat glands become highly active	Increasing danger of heat strokes and cardiovascular embarrassment
35	Hot			
	Warm	Uncomfortable	*Vasomotor regulation against heat*. Blood vessels dilate and allow blood flow as close as possible to the outer surface. Skin temperature increases promoting extra dry heat loss	
30	Slightly warm			
25	Neutral	Comfortable	No action required. Skin temperature decreases reducing dry heat loss	Normal
20	Slightly cool	Slightly uncomfortable	*Vasomotor regulation against cold*. Blood vessels contract preventing flow of blood and transport of heat to outer surface. Outer skin tissues become insulating layer	
	Cool			
15	Cold		Urge for more clothing and exercises	
10	Very cold	Uncomfortable	*Metabolic regulation against cold*. Spontaneous increase in activity Shivering	Increasing complaint from dry mucosa and skin Muscular pain
			Zone of inevitable body cooling	Body unable to combat Impairment of peripheral circulation

Work rates W are difficult to assess although standard values are listed for various types of activity [24, 27].

The total latent heat loss Q_{ev} can be measured in a controlled situation by monitoring body weight. Mass transfer losses consist of the heat content of respired vapour and the latent heat absorbed from the outer surface of the skin by the evaporative process. The relative amounts of metabolic heat dissipated in this way depends upon the degree of physical activity, the level of clothing and its resistance to mass diffusion, the psychrometric properties of the environment, and the ambient air velocity.

The radiant and convective rates of heat transfer Q_r and Q_c depend upon the environmental mean radiant and air temperatures and the surface heat transfer coefficients for radiation and convection.

The presence of clothing makes the transfer paths more complex. Heat is transferred between the surface of the skin and the inner surface of the clothing by conduction, where the body touches the fabric, by gaseous convection and radiation across air cavities and by the mass transfer of liquid and evaporated vapour. This heat is transported through the fabric to the environment by conduction through the fibres and the entrapped air, radiation through the voids, and by the diffusion of liquids and vapours [41].

The Fanger comfort equation [27] is based on the solution of eqn. (3.6) at thermal equilibrium (i.e. $Q_{st} = 0$). Design charts have been produced [27] from which it is possible to predict combinations of environmental parameters which would produce physically comfortable environments for clothed personnel performing selected activities.

3.8 THE ASHRAE COMFORT CHART [24]

The "new" ASHRAE comfort chart has been produced from a combination of studies involving observations of subjective response and of the physiological regulatory processes activated by thermally "uncomfortable" environments. The combined results of this research has been plotted on the coordinates of the psychrometric chart (see Fig. 3.1). The area within the parallelogram represents the ranges of psychrometric conditions for which indoor environments with air velocities less than ~0.25 m s^{-1} and with mean radiant temperatures equal to air dry-bulb temperatures have been rated as being thermally comfortable by sedentary subjects wearing light clothing (i.e. with a thermal resistance of approximately 0.1 m^2 K W^{-1}). A previous comfort zone deduced whilst deriving the "old" effective temperature (see section 3.6.1) is also available [24]. The most commonly recommended inside design condition is where the two zones overlap. This subregion also agrees closely with that recommended by Fanger [27] for the same supplementary conditions.

Psychrometric conditions resulting in uncomfortable thermal environments may be rated using the "new" ASHRAE effective temperature ET^* which is superimposed on the chart at the intersection of lines of constant dry-bulb temperature with the 50% relative humidity curve. Constant ET^* lines are lines of "constant skin wettedness" w which varies between 0 (at $ET^* = 25°C$ for physiological thermal neutrality) to 1.0 (at $ET^* = 41°C$ when the total surface of the skin is wet). Lines of constant skin wettedness were obtained semi-empirically by ASHRAE workers in terms of observed rates of body-weight loss when subjects were exposed to various combinations of environmental factors. Outside the range $25°C < ET^* < 41°C$, the constant ET^* lines are drawn parallel to the lines at 25°C and 41°C respectively. It must be emphasised that, like the "old" effective temperature, ET^* is merely an index of comfort or discomfort, having no mathematical significance in the general heat balance equation. Table 3.2 is provided to aid the interpretation of values of effective temperature.

Chapter 4

Climate

4.1 THE SUN

The sun is a gigantic fusion reactor in which hydrogen is being continuously converted to helium, releasing large quantities of nuclear energy in the process. This results in internal temperatures of the order of $10^7 °C$ and an effective surface temperature (when viewed from the earth) of $\sim 6000 °C$. Solar radiant energy streams out into space at the rate of ~ 70 MW m^{-2} (see section 2.4) at the sun's surface (1.392×10^6 km diameter). The earth (6367 km diameter), situated 149.6×10^6 km distant, intercepts a relatively small fraction ($\sim 10^{11}$ MW or 1.362 kW m^{-2} at the boundary of the upper atmosphere) of the total energy dissipated ($\sim 10^{21}$ MW) [13, 15].

4.2 ELECTROMAGNETIC WAVE SPECTRUM

Because there is no intervening substance, all energy transmission between the sun and the earth takes place by the propagation of electromagnetic waves. These travel through space with a uniform velocity u of 2.9979×10^8 m s^{-1}. Thus activity at the sun is observed at the surface of the earth 500 s later. The energy content of a wave motion may be characterised by its wavelength λ (see Table 2.1);

$$u \times f\lambda \qquad (4.1)$$

where f is the frequency of the wave. Most radiant sources emit energy over a range of wavelengths or frequencies. A single normalised cumulative spectral distribution curve is unique to all blackbody radiant sources (see Fig. 2.14).

The sun may be considered as a blackbody radiator at $5500 °C$ emitting over the waveband 2.9×10^{-7} to 4.75×10^{-6} m (Fig. 4.1). Light is electromagnetic radiation which may be discerned by the visual sensors of man (Figs. 4.2 and 4.3) and is restricted to the range of wavelengths between 3.8×10^{-7} and 7.0×10^{-7} m [42]. The solar radiation incident at the perimeter of the earth's atmosphere contains approximately 5% of ultraviolet radiation, 52% of visible light, and 43% of infrared radiation.

4.3 TERRESTRIAL SOLAR ENERGY INTERCEPTION

When the sun is directly overhead in a cloudless sky, as much direct specular radiation as 1.025 kW m^{-2} can reach the earth's surface. The remaining 0.337 kW m^{-2} is reflected by ideal gas molecules which filter out shorter wavelengths and also selectively transmit light in the blue end of the visible spectrum. The modified radiation contains approximately 1% of ultraviolet radiation, 39% of visible light, and 60% of infrared radiation.

54

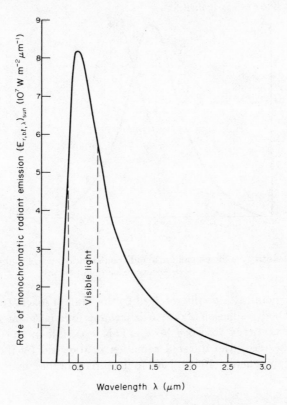

Fig. 4.1. Spectroradiometric curve for the sun.

Solar radiation which has been scattered, or absorbed and reradiated, by a combination of gas molecules, water vapour, and dust particles, becomes diffuse, and the resulting "sky radiation" appears visibly white. Layers of cloud reflect direct and diffuse radiation away from the earth but also insulate against direct radiation losses from the earth's surface. At night, in the absence of cloud cover, up to 6000 W m^{-2} can be lost from the surface of the earth by direct radiation to space. Figure 4.4 illustrates the mean heat balance situation of the earth with respect to the solar energy input and losses to deep space [43]. For the steady-state model shown, each constituent of the diagram is in thermal balance with zero net heat

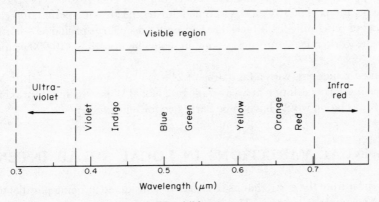

Fig. 4.2. The visible spectrum.

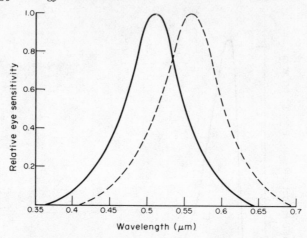

Fig. 4.3. The relative sensitivity of the human eye to radiation at various wavelengths: ——— rod vision (day vision); — — — cone vision (night vision).

gain. All solar energy influx is finally received by the vault of space. It is interesting to note that the total blackbody radiation of the earth accounts for a far greater energy loss (119%) than the rate of solar energy absorbed (47%). This is possible in an equilibrium condition because most of the terrestrial infrared radiation is absorbed by the atmosphere, which reradiates much of this energy (105%) back to the surface of the earth.

4.4 RATE OF ENERGY USAGE

For all practical purposes the sun is the sole primary source of all energy available for consumption on earth. The highest estimate of fossil-fuel reserves (cf. Chapter 1) represents only 40 days of terrestrial solar interception: 6500 times this amount of energy is released by the sun every second! The current rate of fuel consumption corresponds to $\sim 0.001\%$ of the rate of solar energy absorption by the earth.

These figures are, however, no grounds for complacency. Insolation on land masses represents approximately 29% of the total reaching the surface of the earth. Solar radiation falling upon fuel-producing vegetation is a very small fraction of this (perhaps $\sim 0.01\%$). Plants are approximately 1% efficient in converting solar energy to fuel [44]. If a mean rate of solar energy influx of 300 W m^{-2} is assumed during the growing period, it is easily demonstrated that our remaining fossil-fuel store has required a charge period of ~ 65 million years. Each year the world releases an energy content which had taken $\sim 110\,000$ years to accumulate.

The United Kingdom, with a land area of 244 755 km^2 consumes 9×10^{12} MJ annually whilst receiving solar radiation at an average rate of 200 W m^{-2}. Thus approximately 0.5% of total UK insolation would have to be harnessed for self-sufficiency.

4.5 ANNUAL VARIATIONS IN LOCAL SOLAR INTENSITIES

The sun is so far from the earth that its rays may be considered as being parallel to each other where they encounter the earth. The intensity of direct radiation varies from a maximum

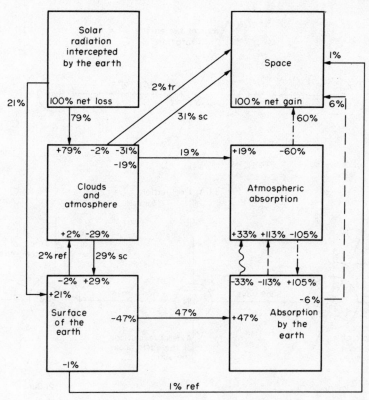

Fig. 4.4. Steady-state energy balance for the earth: ———— solar radiation; ———— terrestrial radiation; ~~~ sensible and latent heat exchanges other than by thermal radiation. ref, reflected; sc, scattered; tr, transmitted.

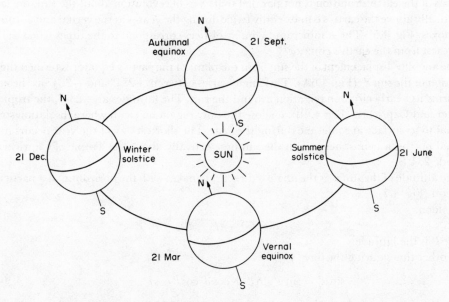

Fig. 4.5. Positions of the earth with respect to the sun at solstices and equinoxes.

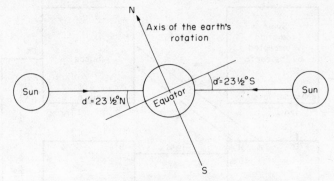

Fig. 4.6A. Declination.

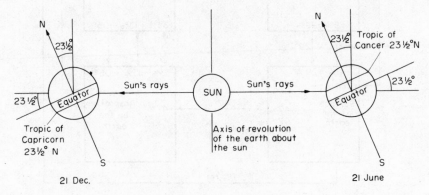

Fig. 4.6B. The tropics.

value near the equator to minimum values near the poles according to a cosine law. Because the axis of the earth's rotation is not parallel to its axis of revolution about the sun, the latter is vertically above the earth's equator only twice during the year—at the vernal and autumnal equinoxes (Fig. 4.5). The summer and winter solstices correspond to the times when the sun is furthest from the earth's equator.

The angular displacement of the sun from the plane of the earth's equator is termed the declination of the sun d' (Fig. 4.6A). This angle varies between $+23\frac{1}{2}°$ and $-23\frac{1}{2}°$ as the earth performs its yearly circumnavigation around the sun. The latitudes at $\pm 23\frac{1}{2}°$, the tropics of Cancer and Capricorn (Fig. 4.6B), enclose the only region on earth where insolation strikes normal to its surface at sometime during the year. The shortest day of the year occurs in the normal northern hemisphere when the sun lies vertically above the Tropic of Capricorn at latitude $23\frac{1}{2}°$S.

The altitude of the sun α' is the angle a direct ray makes with the horizontal at a particular location (Fig. 4.7).

At noon,

$$\alpha' = 90 - (l' - d'), \tag{4.2}$$

where l' is the latitude.

At other times during the day [45],

$$\sin \alpha' = \sin d' \sin l' + \cos d' \cos l' \cos h', \tag{4.3}$$

where h' is the hour angle, or the angular displacement of the sun from its position at noon, i.e.

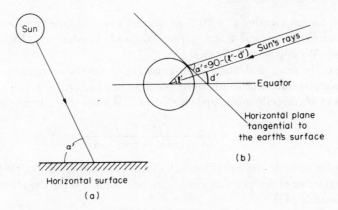

Fig. 4.7. The altitude of the sun: (a) at a horizontal surface; (b) at noon.

$$h' = \frac{360t°}{24},$$ (4.4)

where $t°$ is defined as the "sun-time" in hours before or after the time when the sun is highest in the sky.

Sun-time equals Greenwich Mean Time (GMT) in the United Kingdom. Sun-time for other longitudes can be obtained from Fig. 4.8. This can easily be amended in terms of local times. Thus all solar data for the latitude 51.7°N presented in tables and figures can be

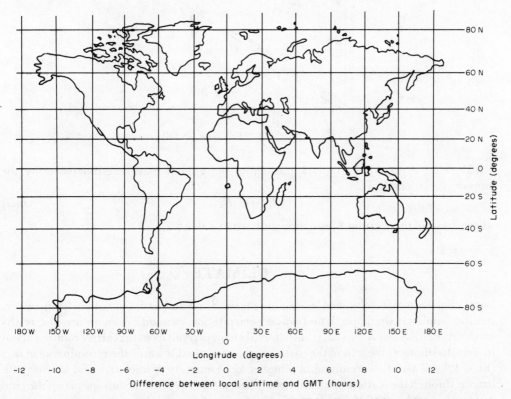

Fig. 4.8. Difference between local sun-time and GMT for all parts of the world.

adapted to apply for any location at $\sim 50°$N longitude. These include London and the border between the United States and Canada (i.e. Portland, Washington, Minneapolis, New York, Vancouver, Winnipeg, and Ottawa).

The azimuth of the sun z' is defined as the angle the horizontal component of a direct ray from the sun makes with the true south in the northern hemisphere (Fig. 4.9).

At noon the sun lies directly south, and so z' is zero. At other times during the day

$$\tan z' = \frac{\sin h'}{\sin h'. \cos h' - \cos l' \tan d'}. \qquad (4.5)$$

Table 4.1 lists solar altitude and azimuth angles for the United Kingdom [19].

The numerical value of direct radiation I_δ may be obtained using the altitude alone from the empirical formula [45]

$$I_\delta = K_1 \exp[-(K_2/\sin \alpha')], \qquad (4.6)$$

where K_1 and K_2 are empirical constants (Table 4.2). Table 4.3 lists calculated values of direct solar intensities normal to the sun and on horizontal and south-facing surfaces [19].

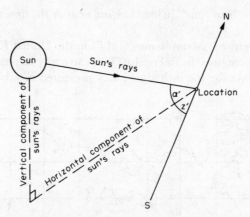

Fig. 4.9. The azimuth angle of the sun.

Values for scattered sky radiation reaching the ground may be estimated using the empirical relationship [45]

$$I_{sc} = K_3 I_\delta, \qquad (4.7)$$

where values of the constant K_3 are also tabulated in Table 4.2.

4.6 CLIMATE

Local climates may be classified in terms of mean values and variations of air temperature, specific humidity, wind speed, and water precipitation. Seasonal changes are due to the earth's revolution about the sun, whilst diurnal changes result from the earth's rotation about its axis. In summer, the axis of the earth's rotation is tilted towards the incoming solar rays (Figs. 4.5–4.7 and 4.10) resulting in a higher local intensity of insolation and shorter path lengths through the scattering atmosphere. Thermal energy from the sun warms up the land surfaces upon which it falls. Some of this energy travels inwards by conduction and is stored

TABLE 4.1. SUN ANGLES FOR LATITUDE 51.7°N

Sun-time	22 Dec.	21 Jan. and 21 Nov.	20 Feb. and 23 Oct.	22 Mar. and 22 Sept.	20 Apr. and 24 Aug.	21 May and 23 July	21 June
(a) Solar altitude angles α'							
0600 and 1800					9	15	18
0700 and 1700			1	10	18	25	27
0800 and 1600		2	10	19	28	34	37
0900 and 1500	6	10	17	27	37	44	46
1000 and 1400	12	15	24	34	44	52	55
1100 and 1300	15	19	28	39	50	58	61
1200	17	20	29	40	51	60	63
(b) Solar azimuth angles z'							
0600				90	83	77	74
0700			108	101	94	88	85
0800		125	120	114	106	100	97
0900	139	138	133	127	120	114	110
1000	152	151	148	143	137	131	128
1100	166	165	163	161	157	153	151
1200	180	180	180	180	180	180	180
1300	194	195	197	199	203	207	209
1400	208	209	212	217	223	229	232
1500	221	222	227	233	240	246	250
1600		234	240	246	254	260	263
1700			252	259	206	272	275
1800				270	277	283	286

TABLE 4.2. CONSTANTS FOR USE IN EQNS. (4.6) AND (4.7)

Month	Jan.	Mar.	May	July	Sept.	Nov.	
K_1	0.058	0.071	0.121	0.136	0.092	0.062	
K_2[a]	1.230	1.186	1.104	1.085	1.152	1.220	*(kW m^{-2})
K_3	0.142	0.156	0.196	0.207	0.177	0.149	

[a] kW m^{-2}

TABLE 4.3. DIRECT SOLAR INTENSITIES I_δ WITH A CLEAR SKY FOR PLACES BETWEEN 0 AND 300 M ABOVE SEA LEVEL (W m^{-2})

Altitude angle	5	10	15	20	25	30	35	40	45	50	60	80
Normal to sun	210	388	524	620	688	740	782	814	840	860	893	920
On horizontal surface	18	67	136	212	290	370	450	523	594	660	773	907
On a south-facing vertical surface	210	382	506	584	624	642	640	624	594	553	477	160

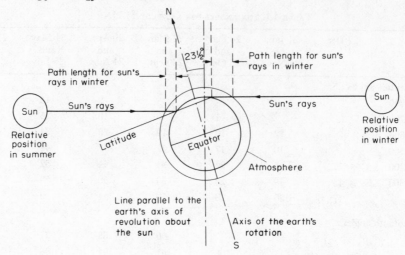

Fig. 4.10. Positions of the sun with respect to the earth in summer and winter for the northern hemisphere.

in the upper layers of the earth's crust. Some is convected to adjacent air layers and some is reradiated to space in long-wave ($\sim 10\ \mu m$) radiation (Fig. 4.4). Thus the geography of a location affects climate, determining how much solar energy is absorbed and stored by the earth, and how readily heat is released to the atmosphere [45]. A balance point is reached at steady state when heat gains equal heat losses. Equilibrium air conditions and surface temperatures of sea and land depend upon the relative amounts of short-wave direct and long-wave indirect radiation absorbed, long-wave radiation reradiated, and heat convected, evaporated, or evapotranspirated to adjacent air layers and to space.

The earth's atmosphere is comparatively transparent to short-wave radiation whilst land masses are opaque and good absorbers. Water is partially transparent and absorbs energy in depth. Since the heat penetrates deeper, surface temperatures of water masses do not reach values as high as those achieved by land surfaces during the daytime. Conversely, the land loses heat more rapidly than the sea at night, resulting in comparatively lower land temperatures. Volumes of water act as thermal buffer stores which smooth out time-dependent variations of solar energy influx. More extreme changes in air temperature occur in the middle of land masses than at coastal locations.

At latitudes greater than 40° the earth loses more net heat by radiation than it receives from the sun, whilst in the equatorial regions, bounded by the latitudes 40°N and 40°S, the earth experiences a net heat gain. The equilibrium process is aided by heat transfers associated with global air and water movements.

4.6.1 Winds

Thermal upcurrents of warm air near the equator and downcurrents of cooler air at higher latitudes cause permanent global air movements (Fig. 4.11) which transfer equatorial heat to colder regions. These give rise to the trade winds, the westerlies, and the doldrums (regions of very low wind speeds). The earth's rotation deviates these bulk air flows (Fig. 4.12) causing regions of high and low pressure—connected by closed vortices—to form. Local characteristics and ocean currents affect the flow patterns further, and so the familiar, but complex, meteorological vortex diagrams are formed.

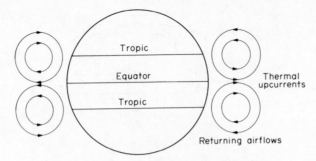

Fig. 4.11. Permanent global air movements.

Local coastal winds are created by air movements resulting from the unequal rates of heating of land and sea. During the day-time, air in contact with the warmer land surfaces receives heat by convection, and expands and rises. The cycle is completed by the downfall of air cooled by the sea (Fig. 4.13). At night, a reversed circulation is established.

The comparative absence of dust particles and the reduction in the amount of water vapour present in the atmosphere at high altitudes mean that radiation outwards to space from high ground is virtually unimpeded. Thus the surfaces of the earth at higher levels cool more rapidly at night. Air layers close to the ground are chilled, become denser, and "slip" down hillsides, forming local gentle "katabatic" or "gravity" winds.

As air rises it expands adiabatically and cools. Water vapour condenses from air when the psychrometric dew-point is reached. Mountain ranges sometimes act, therefore, as natural dehumidifiers when moist warm air is forced by prevailing winds over the peaks (Fig. 4.14). Clouds are formed at the dew-point level and the condensed liquid is precipitated as rain. The dehumidified air descends on the leeward side where it is reheated by warmer lower-level ground surfaces to form a dry and warm local breeze.

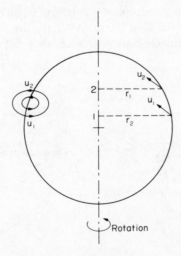

Fig. 4.12. Deviation of global air movements caused by the rotation of the earth about its axis.

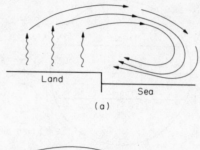

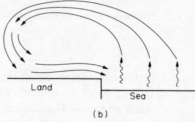

Fig. 4.13. Coastal winds: (a) day-time; (b) night-time.

4.6.2 **Mist, fog, rain, and dew**

The mechanism of water condensation from a psychrometric mixture requires the dry-bulb temperature to fall below the dew-point corresponding to the moisture content, and also the presence of condensation nucleii. These are hygroscopic particles of dirt, salt, air-borne pollutants, or aero-allegens at which condensation can commence. Thus mist and fog patches are commonly encountered over industrial areas where the nucleii are in abundance, or near seas and lakes where the air has a high moisture content. Droplet formation also requires relatively still air, and so wind rapidly disperses fog. "Advection" fog, or sea-mists, form when moist sea breezes blow inland over cooler land surfaces. "Radiation" fog is produced when air in contact with ground, which has lost heat by radiating to the sky, is cooled. Clouds, in impeding radiative losses, inhibit the formation of radiation fog.

Clouds are arrays of minute water droplets which form when moist air rises and cools. The normal fall in temperature is $\sim 0.5°C$ in 100 m. When the droplets become large enough, so that their weights overcome frictional air resistance, they fall as rain. Table 4.4 provides precipitation data.

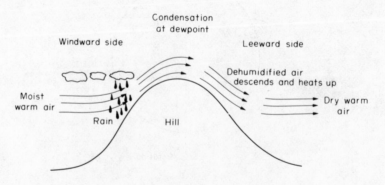

Fig. 4.14. Natural dehumidification.

TABLE 4.4. PRECIPITATION

Location	Precipitation (mm water)		
	Maximum average monthly	Minimum average monthly	Annual mean
Accra	178	15	724
Aden	5	<3	23
Algiers	137	<3	762
Amsterdam	71	33	650
Athens	71	5	401
Baghdad	56	<3	140
Beirut	190	<3	869
Belfast	91	48	856
Berlin	79	33	587
Brussels	89	51	838
Buenos Aires	100	56	950
Cairo	5	<3	28
Calcutta	328	5	1600
Capetown	89	8	508
Chicago	89	51	836
Copenhagen	81	30	592
Dublin	76	48	754
Edinburgh	79	41	701
Entebbe	257	66	1506
Geneva	97	46	861
Hamilton	147	104	1463
Hong Kong	394	30	2162
Kuwait	28	0	130
Lagos	460	25	1836
Lahore	140	3	503
Lisbon	107	5	686
London	64	36	580
Los Angeles	79	3	381
Madrid	56	8	419
Melbourne	66	46	653
Mexico City	170	5	747
Montreal	97	66	1036
Moscow	76	28	630
New York	109	76	1092
Oslo	97	33	683
Ottawa	89	56	871
Paris	51	33	566
Perth	180	8	881
Prague	66	20	490
Reykjavik	102	41	861
Rio de Janeiro	137	41	1082
Rome	97	15	653
San Francisco	119	<3	561
Shanghai	180	35	1135
Singapore	257	170	2413
Sydney	127	71	1181
Tokyo	234	48	1565
Toronto	74	61	815
Tunis	64	3	419
Vancouver	224	30	1458
Vienna	76	36	650
Warsaw	76	28	559
Washington DC	112	66	1247

Surface dew forms when the air in immediate contact with the ground is cooled by gaseous conduction to a temperature below the dew-point. Rates of cooling differ for different solid materials. Rocks have high thermal diffusivities compared with those of soils and grasses. Thus rocks can extract heat from the ground to offset radiative losses more rapidly than can vegetation. The result is that dew forms on grasses and plants before it forms on solid minerals.

4.6.3 **Diurnal variations**

The earth's surface, having radiated to the sky during the night, is coldest just before dawn. It receives maximum solar influx at noon. Because of response lags in the convective transmission between the earth's surface and the surrounding air, the air temperature is highest sometime after noon (2–3 p.m.). The relative humidity of the air near the ground is always 100% when dew is present. Specific humidity is usually measured in meteorological practice twice per day (at 9 a.m. and at 1 p.m.), and it is customary to assume that this does not vary appreciably during that part of the day when the relative humidity is less than 100%. Significant variations in specific humidity occur over longer periods of time with air-bound water movements, evapotranspiration, rainfall (resulting in adiabatic saturation at constant wet-bulb temperature), and local natural and industrial activities (see, typically, Fig. 4.15 for the United Kingdom).

Diurnal dry-bulb temperature and relative humidity fluctuations follow roughly sinusoidal relationships (Fig. 4.16) [45].

4.7 **SUMMER AND WINTER EXTERNAL DESIGN CONDITIONS**

When selecting heating and air-conditioning plant for buildings, it is uneconomic to use absolute maxima and minima external design conditions [19]. When designing insulated

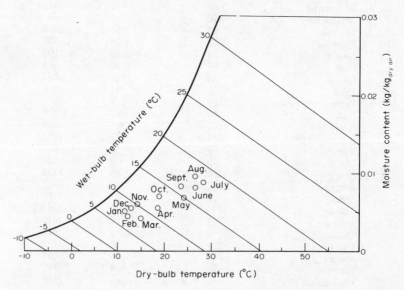

Fig. 4.15. Seasonal variations in air conditions. (Mean monthly maximum conditions for the United Kingdom are shown ○.)

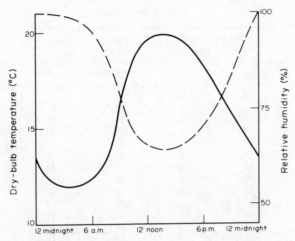

Fig. 4.16. Typical diuranal variation of air conditions: ———— air temperature; — — — relative humidity.

structures, however, the more rigorous application of boundary parameters leads to more efficient thermal units. Figures 4.17A–4.17I and Table 4.5 present summer dry-bulb, summer wet-bulb, and winter dry-bulb design temperatures for the United States and Canada, Europe, and other locations in the world. Summer design dry- and wet-bulb isotherms for the United Kingdom are also shown in Fig. 4.18 and winter design humidities are always assumed to be 100%. Some annual monthly mean dry-bulb temperature variations are listed in Table 4.6 and other climatic data concerning water precipitation, wind speeds, and prevailing wind directions are presented in Table 4.4 and Figs. 4.19 and 4.20 [19].

TABLE 4.5a. DESIGN TEMPERATURES FOR THE UNITED STATES AND CANADA [24]

State	Winter dry-bulb (97½% limit)		Summer dry-bulb (5% limit)		Summer wet-bulb (5% limit)	
	Mean[a]	±[b]	Mean[a]	±[b]	Mean[a]	±[b]
United States						
Alabama	−6	2	34	[c]	25	
Alaska	−30	13	18	5	14	3
Arizona	−4	6	35	4	21	3
Arkansas	−6	2	35		25	
California	2	3	30	8	20	2
Colorado	−17	4	30	4	18	2
Connecticut	−16	2	29		24	
Delaware	−9		32		25	
Florida	4	3	32	2	26	
Georgia	−6	2	34		26	
Hawaii	16		30	3	23	
Idaho	15	3	32	2	17	
Illinois	−17	2	32	1	25	
Indiana	−16	2	32		25	
Iowa	−16	2	31	1	25	
Kansas	−16	1	35	2	24	1
Kentucky	−12	4	33	1	24	
Louisiana	−2	2	34	1	26	
Maine	−21	4	27		21	

Table 4.5a continued

State	Winter dry-bulb (97½% limit)		Summer dry-bulb (5% limit)		Summer wet-bulb (5% limit)	
	Mean[a]	±[b]	Mean[a]	±[b]	Mean[a]	±[b]
Maryland	−10	3	32		24	
Massachusetts	−16	3	28	1	23	
Michigan	−16	2	29	2	23	
Minnisota	−34	3	29	2	22	1
Mississippi	−4	1	34		26	
Missouri	−14	2	34		25	
Montana	−11	3	29	2	17	2
Nebraska	−19	2	34	1	23	2
Nevada	−13	6	33	3	17	2
New Hampshire	−22	3	29		22	
New Jersey	−10	2	31	1	24	
New Mexico	−10	5	33	3	19	2
New York	−16	4	29	1	26	
North Carolina	−6	2	32		26	
North Dakota	−28	1	30	1	21	
Ohio	−15	1	31		23	
Oklahoma	−10	2	36		24	
Oregon	−7	6	29	2	18	1
Pennsylvania	−14	2	28		23	
Rhode Island	−12		28		23	
South Carolina	−4	1	33		25	
South Dakota	−24	2	32	1	22	
Tennessee	−8	1	34	1	25	
Texas	−3	4	36	1	25	1
Utah	−13	5	33	2	18	
Vermont	−23	2	28		21	
Virginia	−9	2	31	2	24	
Washington	−7	5	28	4	17	
West Virginia	−13	2	30	2	24	
Wisconsin	−14	3	29	1	22	
Wyoming	−21	2	29	2	16	1
Canada						
Alberta	−34	3	28	2	18	
British Columbia	−19	13	27	4	18	
Manitoba	−34	3	27	2	21	
New Brunswick	−24	2	27	2	21	
Newfoundland	−21	7	25	1	18	
Northwest Territories	−43	1	20	8	17	
Nova Scotia	−18	3	25	2	19	
Ontario	−22	6	29	1	22	1
Quebec	−26	3	27	2	21	
Saskatchewan	−34	2	29	1	21	
Yukon Territory	−41		22		15	

[a] Arithmetic mean values indicated by the various weather stations within the State. Figures are rounded to nearest whole numbers.

[b] Standard deviation.

[c] Blank space indicates a standard deviation which is less than unity.

TABLE 4.5b. DESIGN TEMPERATURES AT VARIOUS GLOBAL LOCATIONS [24]

Country	Winter dry-bulb (97½% limit)		Summer dry-bulb (5% limit)		Summer wet-bulb (5% limit)	
	Mean[a]	±[b]	Mean[a]	±[b]	Mean[a]	±[b]
Aden	21		37		28	
Algeria	−7		32		24	
Argentina	0		33	3	24	
Australia	7	5	32	4	23	3
Austria	−11		28		19	
Azores	9		25		22	
Bahamas	17		32		26	
Belgium	−7		25		19	
Bermuda	13		33		26	
Bolivia	0		20		13	
Brazil	14	3	30	2	25	
British Honduras	16		32		27	
Bulgaria	−13		33		20	
Burma	15	3	31		27	
Cambodia	20		34		28	
Ceylon	20		32		27	
Chile	2	5	25	7	17	4
China	0		34		26	
Columbia	14	6	27	6	21	5
Congo	18		32		27	
Cuba	18		33		27	
Czechoslovakia	−13		28		18	
Denmark	−7		23		18	
Dominican Republic	18		31		27	
Ecuador	10		32		21	7
El Salvador	13		35		24	
Ethiopia	5		27		17	
Finland	−18		22		17	
France	−6	5	28	2	20	1
French Guiana	22		33		28	
Germany	−10	3	25	2	19	
Ghana	19		32		26	
Gibraltar	6		30		23	
Greece	1		33		23	2
Greenland	−22		16		11	
Guatemala	10		27		19	
Guyana	23		30		26	
Haiti	19		34		27	
Honduras	10		29		22	
Hong Kong	10		33		27	
Hungary	−10		29		21	
Iceland	−8		13		12	
India	14	5	38	3	27	1
Indonesia	21	1	32	1	26	
Iran	−3	8	38	5	23	4
Iraq	1		43		22	
Ireland	−4		21		17	
Israel	4		33		31	
Italy	−1	4	30	1	23	
Ivory Coast	20		31		27	
Japan	−6		30		25	
Jordan	2		33		20	
Kenya	10		26		18	
Korea	−14		30		25	
Lebanon	7		32		24	
Liberia	20		31		27	
Libya	9		32		24	

Table 4.5b continued

Country	Winter dry-bulb (97½% limit)		Summer dry-bulb (5% limit)		Summer wet-bulb (5% limit)	
	Mean[a]	±[b]	Mean[a]	±[b]	Mean[a]	±[b]
Madagascar	9		29		22	
Malaysia	22		32		27	
Martinique	19		31		27	
Mexico	9	6	32	3	22	5
Morocco	5		30		21	
Nepal	1		30		24	
Netherlands	−5		22		17	
New Guinea	22		30		27	
New Zealand	2	3	24	1	18	
Nicaragua	19		33		26	
Nigeria	20		32		27	
Norway	−11		23		18	
Pakistan	7	5	37	4	27	
Panama	23		33		27	
Paraguay	8		36		27	
Peru	13		29		23	
Philippines	23		33		27	
Poland	−14		25		19	
Portugal	4		28		19	
Puerto Rico	20		30		26	
Romania	20		32		21	
Saudi Arabia	10		41		27	2
Senegal	16		33		27	
Somalia	20		32		27	
South Africa	3		28		21	
Soviet Union	−20	6	26	4	18	2
Spain	1	3	31		21	
Sudan	13		40		24	
Surinam	20		33		27	
Sweden	−13		22		16	
Switzerland	−10		26		19	
Syria	0		37		21	
Taiwan	9		33		27	
Tanzania	18		31		27	
Thailand	17		34		27	
Trinidad	18		31		26	
Tunisia	5		35		22	
Turkey	−3	5	33	2	23	3
United Arab Republic	8		37		24	
United Kingdom	−3	1	22	2	17	
Uruguay	4		29		22	
Venezuela	18		31		24	
Vietnam	14		34		29	
Yugoslavia	−10		29		23	

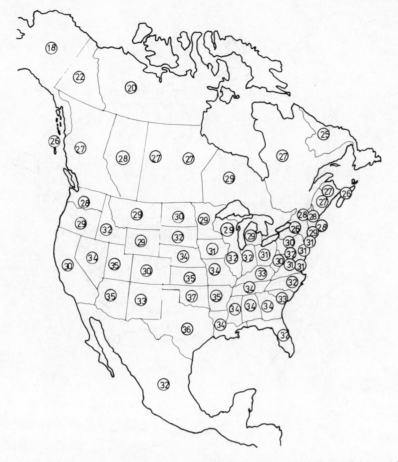

Fig. 4.17A. Summer high (5% limit) design dry-bulb temperatures (°C) for the United States and Canada.

4.8 MICROCLIMATE

The urbanisation of a previously rural area promotes changes in local climate. The relatively rough surface presented to the sky by a city skyline is thermally blacker than surrounding country areas. Thus solar radiation is absorbed more easily by built-up areas than by underdeveloped regions. Air flows are made more turbulent by building conurbations, although average wind speeds can be retarded by up to 25% [46]. Smoke, waste gases, and pollutants above the city skyline impede longwave radiative heat loss and also encourage the formation of fog and rain. Local energy release increases local air temperatures. All these effects can lead to internal air temperatures which are 1–3°C above those encountered in the surrounding countryside. In addition, drainage, cooling towers, and increased water usage result in higher specific humidities. Warmed air rises over cities, and cold air converges on to the built-up area to make good the deficit [47]. The flow patterns created contain a "heat island" which tends to prevent the dispersion of air pollutants (Fig. 4.21).

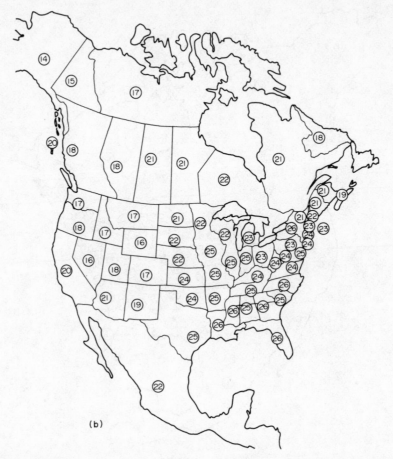

(b)

Fig. 4.17B. Summer high (5% limit) design wet-bulb temperatures (°C) for the United States and Canada.

4.9 LARGE-SCALE WEATHER MODIFICATION

The combustion of fossil fuels has significantly increased the proportion of carbon dioxide present in the earth's atmosphere [47]. Although this has little effect on the amount of solar radiation reaching the earth's surface, the carbon dioxide balance affects thermal equilibrium conditions by inhibiting long-wave radiation transmission. The transparency of the atmosphere is also decreased by the use of aerosols and improper agricultural practices. The global heat balance is also being modified by changing surface characteristics due to widespread urbanisation, deforestation, agricultural activity, the formation of artificial inland lakes, and the deposition of thin oil layers on water surfaces. All fossilised energy released eventually appears as heat which must be dispersed to outer space to maintain established steady-state temperatures. Present knowledge is insufficient to evaluate the aggregated effects of these many factors on world climate. It is, however, commonly agreed [9, 15, 47] that uncurbed industrial growth and associated increased energy consumption will lead to either a 100 m rise in mean sea-level arising from glacial melting, or, alternatively, a drastic decrease in mean air temperatures caused by a reduction of insolation.

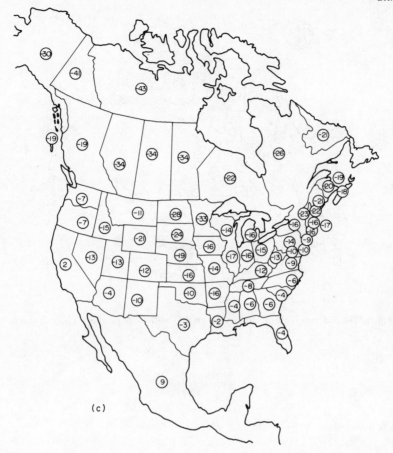

Fig. 4.17C. Winter low (97½% limit) design dry-bulb temperatures (°C) for the United States and Canada.

4.10 THE HEAT LIMIT

The maximum rate of thermal energy rejection by mankind is restricted by a "heat limit" [15] when mean air temperatures, local climates, and ecological cycles would be seriously affected. It has been suggested [15] that this will occur when the rate of world energy consumption approaches $\sim 1\%$ of the rate of absorption of solar radiation by the earth and its atmosphere (i.e. $\sim 10^{10}$ MW). Current world consumption is $\sim 1.3 \times 10^6$ MW (cf. Table 1.1). If mankind were to harness a virtually unlimited source of energy, and if present rates of increased usage were to continue, the heat limit would be reached in approximately 170 years time. The figure of 1% is quite arbitrary, and since the corresponding level of energy release would lead to an increase in mean global temperature of $\sim 6°C$, the acceptable maximum should probably be set lower. Current rates of energy dissipation are responsible for an average rise in world air temperature of $\sim 0.002°C$. Thermal pollution is already a cause for concern on a local level. The United Kingdom contains pockets where energy is rejected to the atmosphere at rates approaching 3.0 W m^{-2}—1.5% of the mean UK solar influx. It is clear that restraint must be exercised by the developed nations, and that local and global climatic variations must be very closely monitored.

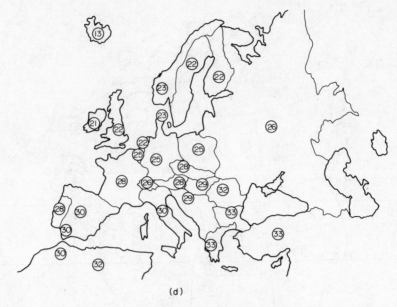

(d)

Fig. 4.17D. Summer high (5% limit) design dry-bulb temperatures (°C) for Europe.

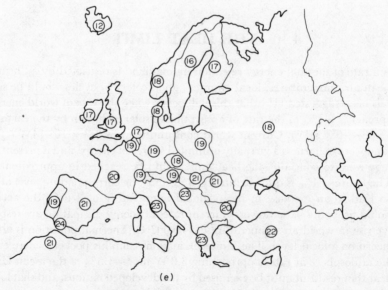

(e)

Fig. 4.17E. Summer high (5% limit) design wet-bulb temperatures (°C) for Europe.

Fig. 4.17F. Winter low (97½% limit) design dry-bulb temperatures (°C) for Europe.

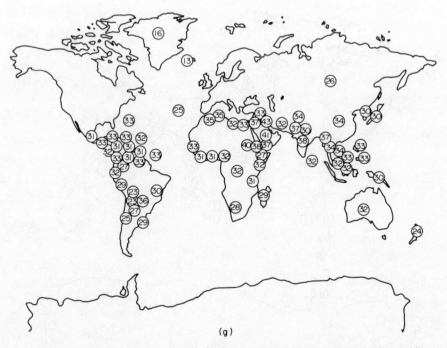

Fig. 4.17G. Summer high (5% limit) design dry-bulb temperatures (°C) for the world.

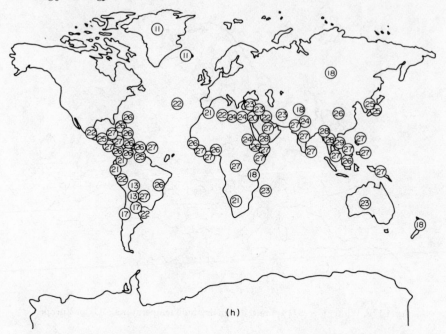

Fig. 4.17H. Summer high (5% limit) design wet-bulb temperatures (°C) for the world.

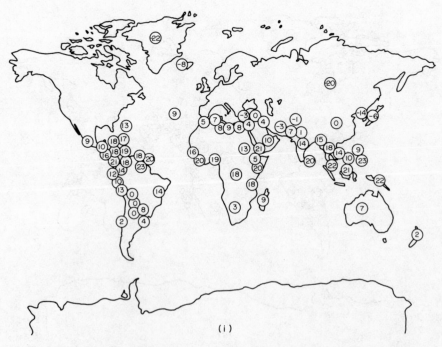

Fig. 4.17I. Winter low (97½% limit) design dry-bulb temperatures (°C) for the world.

(a)

Fig. 4.18A. Summer high design dry-bulb isotherms (°C) for the United Kingdom (exceeded 1% of the months from June to September inclusive).

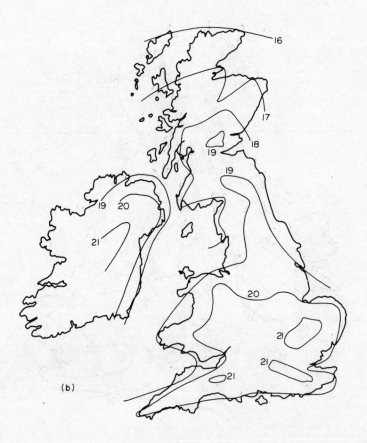

Fig. 4.18B. Summer high design wet-bulb isotherms (°C) for the United Kingdom (exceeded 1% of the months from June to September inclusive).

Fig. 4.19. Basic wind speeds (m s⁻¹) in the United Kingdom. The contours represent maximum gust speeds which are likely to be exceeded only once in 50 years at 10 m above the ground in open level country.

Fig. 4.20. Wind rosettes for the United Kingdom.

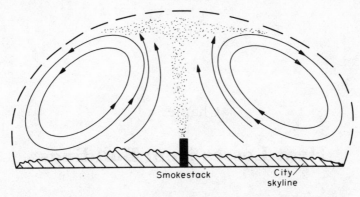

Fig.4.21. Self-contained dust dome over a large industrial city showing how air circulation patterns prevent pollutant dispersion.

TABLE 4.6. MEAN MONTHLY TEMPERATURES (°C)

Place	J	F	M	A	M	J	J	A	S	O	N	D	Mean
United States													
New York	0	1	4	10	15	20	23	22	19	12	7	0	11
New Orleans	13	16	17	21	24	27	27	27	25	21	16	10	20
San Diego	12	12	12	14	16	17	19	20	20	18	16	13	16
Sacramento	7	9	11	13	17	21	22	22	21	18	14	7	15
Denver	−1	0	4	8	13	18	21	19	16	10	4	0	9
Canada													
Edmonton	−13	−10	−5	4	8	13	16	16	10	4	−4	−10	2
Montreal	−10	−8	−3	4	13	18	20	19	15	7	0	−6	6
St. Johns	−6	−6	0	4	7	10	15	16	10	4	−4	−10	2
Europe													
Athens	8	8	12	16	18	23	27	27	22	20	14	11	17
Bergen	0	0	3	7	10	12	16	13	11	9	5	0	7
Berlin	0	0	5	8	13	18	20	19	16	10	4	0	9
Bordeaux	6	6	7	10	13	18	20	20	16	13	10	5	12
Lisbon	10	11	13	16	17	20	22	21	20	17	15	11	16
London	4	4	7	9	13	16	19	17	14	10	7	5	11
Moscow	−12	−10	−3	3	10	16	20	16	11	5	−3	−10	4
Palermo	12	10	13	16	17	21	23	24	20	18	16	12	17
Warsaw	−5	−2	2	8	12	17	18	17	15	10	4	−1	8
Rest of the World													
Aden	26	26	27	28	29	31	30	29	29	28	25	24	28
Algiers	8	8	13	15	16	21	23	24	23	19	15	12	16
Baghdad	10	10	15	21	26	30	34	33	30	24	16	8	21
Buenos Aires	22	22	21	16	10	8	8	8	11	16	18	21	15
Calcutta	19	21	27	28	29	29	28	28	28	26	22	19	25
Capetown	21	21	21	17	15	12	11	12	13	16	17	20	16
Darwin	28	29	29	27	26	25	24	26	27	29	29	29	27
Entebbe	21	21	21	21	21	21	21	21	21	21	21	21	21
Melbourne	19	19	18	16	12	10	10	8	11	16	10	5	14
Tokyo	2	5	7	12	17	21	25	24	21	16	10	5	14

Chapter 5

Heat Losses from Buildings

5.1 INTRODUCTION

Twenty-eight per cent of the primary fuel combusted for consumption in the United Kingdom is used for heating buildings. This represents an energy release of $\sim 2.47 \times 10^{12}$ MJ per year, or $\sim 7.82 \times 10^{7}$ kW equivalent continuous dissipation. Buildings must be heated in order to make up thermal losses by transmission through the structure together with the load induced by ventilation requirements. The analysis of transmission losses is not complex but it is tedious, requiring iterative procedures. In order to illustrate the analytical technique, a representative cubical structure of side 10 m will be examined (Fig. 5.1). Exact solutions of building heat balance should strictly consider the transient behaviour of the system with respect to varying external air temperature, solar influx, and intermittent internal requirements. An estimate of total annual losses in the United Kingdom may, however, be obtained by adopting an annual mean outside environmental temperature of 11°C (at 51.7°N) (Fig. 5.2 and Table 5.1). Then, with a constant inside temperature of 25°C, the average temperature excess is 14°C. The transmission heat loss from a building is calculated from

$$Q_{tr} = \sum UA\Delta T. \tag{2.2}$$

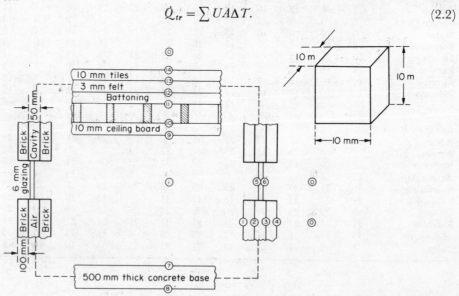

Fig. 5.1. Sketch detailing the structure analysed.

82

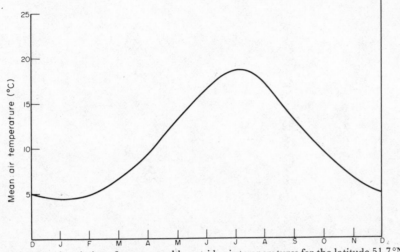

Fig. 5.2. Annual variation of mean monthly outside air temperatures for the latitude 51.7°N.

TABLE 5.1. MONTHLY AVERAGE OUTSIDE AIR TEMPERATURE
(51.7°N) [19]

Month	Mean air temperature (°C)
January	4.0
February	4.5
March	7.0
April	9.0
May	13.0
June	16.0
July	19.0
August	17.0
September	14.0
October	10.0
November	6.5
December	5.0
Yearly average	11.0

The overall thermal conductance of the building depends upon the conductances of its components.

5.2 CONSTRUCTION DETAILS

Particulars of the structure under consideration are given in Table 5.2 and the thermal network (Fig. 5.3) illustrates the thermal paths.

5.3 COMPONENT RESISTANCES AND THE OVERALL HEAT TRANSFER COEFFICIENT

The estimation of the rate of steady-state heat loss from the structure requires the evaluation to each of the resistances $R_{i,1}$ to $R_{14,o}$, and the temperatures at each location designated in

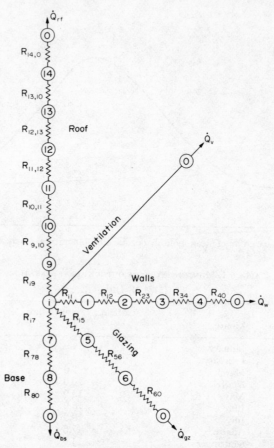

Fig. 5.3. Equivalent thermal network for the structure described in Fig. 5.1.

TABLE 5.2. DETAILS OF THE STRUCTURE CONSIDERED

Height: 10 m
Roof: flat
Glazing: 30% of vertical walling

Component	Materials	Thickness (mm)	Material conductivity ($W\ m^{-1}\ K^{-1}$)	Area (m^2)
Base	Solid concrete	500	1.10	100
Walls	Brick Unventilated air-filled cavity	100/50/100	0.3	280
Glazing	Single glass	6	0.76	120
Roof	10 mm tiles on battens 3 mm roofing felt Rafters (15 mm deep) 10 mm ceiling board	—	see Appendix IV	100

Fig. 5.3. The heat loss through each component is then inversely proportional to the sum of the resistances present between the room air at a mean temperature T_i ($=25°C$) and the outside environment at a mean temperature of T_o ($=11°C$).

Heat is transferred between two fluids across a solid boundary by a three-step steady-state process: from the warmer fluid to the solid surface by convection; through the wall by solid conduction; and by convection from the cooler surface to the surrounding fluid. The heat transferred over an area A (m²) is given by

$$Q = UA\Delta T, \tag{5.1}$$

where ΔT (K) is the difference between the bulk fluid temperatures either side of the wall and U is an overall heat transfer coefficient (W m⁻² K⁻¹) defined by

$$\frac{1}{U} = \frac{1}{h_1} + \frac{\delta}{k_s} + \frac{1}{h_2}. \tag{5.2}$$

The magnitudes of the partial convective heat transfer coefficients h_1 and h_2, depend upon the transport properties and the flow states of the adjacent fluids. k_s and x are the thermal conductivity (W m⁻¹ K⁻¹) and thickness (m) of the solid wall measured in a direction perpendicular to the direction of heat flow.

The total thermal resistance of the section is the reciprocal of U, whilst partial resistances are the reciprocals of the individual terms in eqn. (5.2). Partial heat transfer coefficients are often stated in a dimensionless form using the Nusselt number Nu (section 2.5.2),

$$Nu = \frac{hL}{k_f}, \tag{2.31}$$

where k_f is the fluid conductivity (W m⁻¹ K⁻¹).

Nusselt numbers for free convective flows are functions of the surface-to-fluid Grashof numbers Gr,

$$Gr = \frac{\rho^2 g\beta\Delta TL^3}{\mu^2} \tag{2.38}$$

and the fluid Prandtl number Pr,

$$Pr = \frac{\nu}{\alpha}, \tag{2.33}$$

where the symbols have the meanings as defined previously. The natural convective flow can be laminar or turbulent.

5.4 SIMPLE RELATIONSHIPS FOR HEAT TRANSFER COEFFICIENTS IN AIR

Pertinent generalised Nusselt number relationships have been listed previously (section 2.5.4). Because the fluid under consideration is atmospheric air (assumed dry) at a mean temperature of 18°C, these equations may be simplified as follows:

Values of primary quantities for dry air at atmospheric pressure and at 18°C
   Density ρ, 1.24 kg m⁻³
   Thermal conductivity k, 0.025 W m⁻¹ K⁻¹
   Specific heat c_p, 1004 J kg⁻¹ K⁻¹

Dynamic viscosity μ, 17×10^{-6} kg m^{-1} s^{-1}

Coefficient of volumetric expansion β, 0.0034 K^{-1}

Values of derived quantities for dry air at atmospheric pressure and at 18°C

Specific heat capacity c_p, 1240 J m^{-3} K^{-1}

Thermal diffusivity α, 20×10^{-6} m^2 s^{-1}

Kinematic viscosity ν, 14.2×10^{-6} m^2 s^{-1}

Prandtl number Pr, 0.71

$\dfrac{g\beta}{\nu^2}$, 165×10^6 m^{-3} K^{-1}

$Nu_x = 40 \, h_x x$ $\overline{Nu}_L = 40 \, \bar{h}_L L$

$Gr_x = 1.7 \times 10^8 \Delta Tx^3$ $Gr_L = 1.7 \times 10^8 \Delta TL^3$

FREE CONVECTION

  Vertical flat plate   *Laminar flow*

(2.52)→
$$h_x = 1.065 \left(\frac{\Delta T}{x}\right)^{0.25}$$
(5.3)

(2.53)→
$$\bar{h}_L = 1.37 \left(\frac{\Delta T}{L}\right)^{0.25},$$
(5.4)

with transition to turbulence when $\Delta TL^3 \simeq 6$.

  Vertical flat plate   *Turbulent flow*

(2.59)→
$$h_x = 2.1 \, \Delta T^{0.33},$$
(5.5)

(2.60)→
$$\bar{h}_L = 1.75 \, \Delta T^{0.33}.$$
(5.6)

If the plate is inclined at an angle ψ to the vertical,

$$Gr_x = 1.7 \times 10^8 \Delta Tx^3 \cos \psi$$

  Horizontal plate   *Laminar flow*

Heated plate facing upwards or cooled plate facing downwards,

(2.54)→
$$\bar{h}_L = 1.4 \left(\frac{\Delta T}{L}\right)^{0.25}$$
(5.6a)

(L is the dimension of a side.)

Heated plate facing downwards or cooled plate facing upwards,

(2.55)→
$$\bar{h}_L = 0.7 \left(\frac{\Delta T}{L}\right)^{0.25},$$
(5.7)

with transition to turbulence when $\Delta TL^3 \simeq 200$.

  Horizontal plate   *Turbulent flow*

Heated plate facing upwards or cooled plate facing downwards,

(2.61)→
$$\bar{h}_L = 1.88 \Delta T^{0.33}.$$
(5.8)

Vertical enclosed air spaces   *Laminar flow*

When $\Delta T \delta^3 < 10^{-5}$, natural convection is suppressed and conduction through the air controls (δ is the width of the cavity). When $T\delta^3 > 10^{-5}$,

$(2.56) \rightarrow$
$$h_\delta = 0.51 \left(\frac{\Delta T}{\delta}\right)^{0.25} \left(\frac{L}{\delta}\right)^{-0.111}, \qquad (5.9)$$

with transition to turbulence when $\Delta T\delta^3 \simeq 2 \times 10^{-3}$.

Vertical enclosed air spaces   *Turbulent flow*

$(2.62) \rightarrow$
$$h_\delta = 0.96 \Delta T^{0.33} \left(\frac{L}{\delta}\right)^{-0.111} : \qquad (5.10)$$

Horizontal enclosed air spaces   *Laminar flow*

When $\Delta T\delta^3 < 5 \times 10^{-5}$ natural convection is suppressed and conduction through the air controls.

When $\Delta T\delta^3 > 5 \times 10^{-5}$,

$(2.57) \rightarrow$
$$h_\delta = 0.55 \left(\frac{\Delta T}{\delta}\right)^{0.25}, \qquad (5.11)$$

with transition to turbulence when $T\delta^3 \simeq 2 \times 10^{-3}$.

Horizontal enclosed air spaces   *Turbulent flow*

$(2.63) \rightarrow$
$$h_\delta = 1.01 \left(\frac{\Delta T}{\delta}\right)^{0.33} : \qquad (5.12)$$

5.5   LOCAL TEMPERATURES AND COMPONENT RESISTANCES

From Figs. 5.1 and 5.3:

Designation	Location	Temperature (°C)
i	Inside the building	25
1	At the inside surface of the wall	T_1
2	At the inner face of the air cavity	T_2
3	At the outer face of the air cavity	T_3
4	At the outside surface of the wall	T_4
5	At the inside surface of the glazing	T_5
6	At the outside surface of the glazing	T_6
7	At the inside surface of the base	T_7
8	At the outer surface of the base	T_8
9	At the inner surface of the ceiling	T_9
10	At the outer surface of the ceiling board	T_{10}
11	At the inner surface of the roofing felt	T_{11}
12		T_{11}
13	At the contact between the felt and the tiles	T_{13}
14	At the outer surface of the roof	T_{14}
0	Environmental air	11

Resistance $(m^2 \, K \, W^{-1})$	Type	Applicable equations
R_{i1}	Surface convective and radiative	(5.3)–(5.6)
R_{12}	Solid conduction through the brick	(2.86)
R_{23}	Cavity convective and radiative	(5.9) or (5.10)
R_{34}	Solid conduction through the brick	(2.86)
R_{4o}	Surface convective and radiative	(5.3)–(5.6)
R_{i5}	Surface convective	(5.3)–(5.6)
R_{56}	Solid conduction through the glass	(2.86)
R_{6o}	Surface convective	(5.3)–(5.6)
R_{i7}	Surface convective	(5.7)
R_{78}	Solid conduction through the concrete	(2.86)
R_{8o}	Solid conduction through the earth	(see text)
R_{i9}	Surface convective	(5.6) or (5.8)
$R_{9,10}$	Solid conduction through ceiling board	(2.86)
$R_{10,11}$	Cavity convective and radiative	(5.11) or (5.12)
$R_{11,12}$	Solid conduction through the battens	(2.86)
$R_{12,13}$	Solid conduction through the felt	(2.86)
$R_{13,14}$	Solid conduction through the tiles	(2.86)
$R_{14,o}$	Surface convective and radiative	(5.6) or (5.8)

$\dot{Q}_w$ (W) is the rate of heat loss through the walls
$\dot{Q}_{gz}$ (W) is the rate of heat loss through the glazing
$\dot{Q}_{bs}$ (W) is the rate of heat loss through the base
$\dot{Q}_{rf}$ (W) is the rate of heat loss through the roof

It is necessary to evaluate the surface temperatures, T_1, T_2, T_3, T_4, T_5, T_6, T_7, T_9, T_{10}, T_{11}, and T_{14} because the free convective relationships used to estimate the surface heat transfer coefficients (and hence the boundary and cavity resistances, R_{i1}, R_{23}, R_{4o}, R_{i5}, R_{6o}, R_{i7}, R_{i9}, $R_{10,11}$, and $R_{14,o}$) contain the temperature differences between the surfaces and the adjacent air (except in the case of cavity heat flow when ΔT represents the surface-to-surface temperature difference). Thus the evaluation of the heat loss through each component requires an initial estimate of these temperatures followed by iteration. The procedure will be demonstrated in some detail for the heat transmission through the cavity wall.

5.6 HEAT TRANSFER THROUGH A CAVITY WALL

The building under consideration has 280 m² of vertical cavity walling comprising two 100 mm thick brick walls separated by a 50 mm thick unventilated air gap (see Fig. 5.1). The overall thermal resistance of unit area of the structure is given by

$$R_w = R_{i1} + R_{12} + R_{23} + R_{34} + R_{4o}, \tag{5.13}$$

where

$$R_{i1} = (\bar{h}_{i1})^{-1} \quad \text{and} \quad R_{4o} = (\bar{h}_{4o})^{-1}$$

and

$$R_{12} = R_{34} = \delta/k.$$

Initial rough estimates of the surface temperatures are necessary to evaluate the transport properties, the Grashof numbers, and hence the heat transfer coefficients. The assumption of mean applicable relationships (5.3)–(5.12) within the temperature range 11–25°C leads to an error of less than 10% in the Grashof numbers and less than 5% in the calculated values of the partial heat transfer coefficients because the Grashof number is raised to fractional powers in the equations. These small errors are further filtered when eqn. (5.13) is applied. The numerical values of the temperature differences are, however, more critical, and hence iterative procedures are still necessary.

Initial temperatures are assumed as follows:

$$T_i = 25°C \qquad T_2 = 20°C \qquad T_4 = 13°C$$

$$T_1 = 23°C \qquad T_3 = 16°C \qquad T_o = 11°C$$

Inside partial heat transfer coefficient $\bar{h}_{i1}$. It is first necessary to ascertain the point of transition to turbulence in the surface-bound air layer, i.e. when $(25 - T_1) L^3 = 6$ or $L = 1.44$ m from the base of the wall.

Equation (5.4) applies for the laminar region,

$$\bar{h}_{i1,lam} = 1.09 \text{ W m}^{-2} \text{ K}^{-1},$$

and eqn. (5.6) applies for the turbulent portion,

$$\bar{h}_{i1,tb} = 2.21 \text{ W m}^{-2} \text{ K}^{-1}.$$

The mean inside partial heat transfer coefficient is then obtained by a standard weighting procedure:

$$\bar{h}_{i1} = \frac{h_{i1,lam}L_{lam} + h_{i1,tb}}{L} \qquad (5.14)$$

$$= 2.07 \text{ W m}^{-2} \text{ K}^{-1}$$

and so
$$R_{i1} = 0.48 \text{ m}^2 \text{ K W}^{-1}.$$

Outside partial heat transfer coefficient $\bar{h}_{4o}$. Transition to turbulence occurs at about 1.44 m from the top of the outside wall,

$$\bar{h}_{4o} = 2.07 \text{ W m}^{-2} \text{ K}^{-1} \qquad \text{and} \qquad R_{4o} = 0.48 \text{ m}^2 \text{ K W}^{-1}.$$

Cavity resistance R_{23}. Turbulence would occur in the cavity if its width were greater than ~ 1.15 m. Thus laminar flow prevails across the 50 mm air gap. From eqn. (5.9),

$$\bar{h}_{23} = 0.63 \text{ W m}^{-2} \text{ K}^{-1} \qquad \text{and} \qquad R_{23} = 1.59 \text{ m}^2 \text{ K W}^{-1}.$$

Solid wall resistance R_{12} ($= R_{34}$).

$$R_{12} = R_{34} = 0.1/0.3 = 0.33 \text{ m}^2 \text{ K W}^{-1}.$$

Total resistance to heat flow across the wall section R_w. From eqn. (5.13),

$$R_w = 3.21 \text{ m}^2 \text{ K W}^{-1} \qquad \text{and} \qquad U_w = 0.31 \text{ W m}^{-2} \text{ K}^{-1},$$

and so the first estimate of the *heat transmission* through the wall yields

$$\dot{Q}_w = U_w A(T_i - T_o)$$

$$= 1.2 \text{ kW}.$$

Check on initial temperature estimates. After the first iteration, from potentiometric ratios,

$$T_i = 25°C \qquad T_3 = 16.5°C$$
$$T_1 = 23°C \qquad T_4 = 13°C$$
$$T_2 = 19.5°C \qquad T_o = 11°C$$

These values are sufficiently close to the initial estimates to avoid the necessity for further iteration.

It is interesting to note that the resistance of the solid brick (0.66 m² K W⁻¹) contributes only 20% to the resistance of the whole.

Modifications for the effects of wind speeds and radiative interchanges. A high wind speed ($>15 \,\mathrm{m\,s^{-1}}$) together with external radiative losses, can effectively reduce the outside resistance R_{4o} to negligible proportions. The total resistance of the wall then becomes 2.73 m² K W⁻¹ and the overall heat transfer coefficient U_o increases to 0.37 W m⁻² K⁻¹. Because internal air movements are typically of the order of 0.1–0.3 m s⁻¹, the internal heat transfer coefficient h_{i1} will not diverge appreciably from ~ 2 W m⁻² K⁻¹. Net radiative interchanges inside the building are zero when the temperatures of the inside surfaces are equal. The resistance of the cavity, however, is approximately halved by radiative heat transfer. Bearing these effects in mind and performing a further iterative calculation to match component resistances and temperature drops, the final figures become:

$$R_{i1} = 0.4 \text{ m}^2 \text{ K W}^{-1}, \qquad R_w = 2.08 \text{ m}^2 \text{ K W}^{-1},$$
$$R_{12} = R_{34} = 0.33 \text{ m}^2 \text{ K W}^{-1}, \qquad U_w = 0.48 \text{ W m}^{-2} \text{ K}^{-1},$$
$$R_{23} = 1.02 \text{ m}^2 \text{ K W}^{-1}, \qquad Q_w = 1.88 \text{ kW}.$$

5.7 HEAT LOSSES THROUGH GLAZING, BASE, AND ROOF

Glazing. Most commercial glasses transmit radiation over the wavelength range 0.35–2.7 μm and are opaque at all other wavelengths. An examination of the spectral distribution data for surfaces at 25°C shows that less than 0.1% of this quality radiant energy is emitted and transmitted through windows. Thus the resistance to transmission losses through glazing may be estimated by considering partial inside surface and solid resistance alone.

Base. It is generally recommended [19] that the ex-system temperature to be adopted in calculations regarding heat losses through the ground floors of buildings should be of the order of 10°C above that of the ambient air (i.e. 21°C in the system considered here). The overall thermal resistance then comprises the inside surface resistance (calculated using eqn. (5.7)) and the solid bulk resistance of the base material.

Roof. An expression for the overall thermal conductance of a pitched roof in terms of its components is derived in the appendix to this chapter. The structure presently under investigation has a flat roof consisting of 10 mm tiles with 3 mm of roofing felt on battens affixed to 150 mm deep rafters in parallel with 10 mm of ceiling board (see Fig. 5.1). The resistance of the tile battens and the effects of the rafters may be neglected because the major opposition to heat flow occurs in the inter-cavity air layer.

Summary of losses. Figure 5.4 and Table 5.3 summarises the values of partial resistances, local temperatures, and heat flows through each component, calculated using the iterative technique demonstrated earlier.

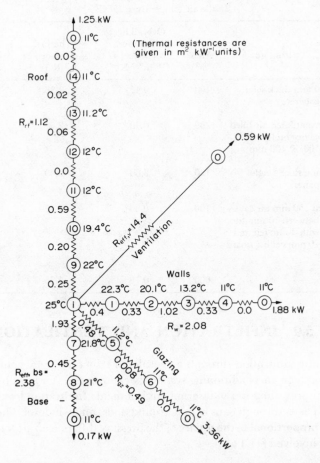

Fig. 5.4. Summary of calculated thermal resistances ($m^2\ K\ W^{-1}$), local temperatures (°C), and heat losses (kW) for the structure described in Fig. 5.1.

5.8 OVERALL *U*-VALUE

An effective overall thermal conductance for a building can be defined by

$$U = \frac{\sum Q}{\sum A \, \Delta T},$$ (5.15)

where $\sum A$ is the total inside surface area.

For the structure considered,

$$U = 0.79 \text{ W m}^{-2} \text{ K}^{-1}$$

and the overall effective thermal resistance

$$R = 1.26 \text{ m}^2 \text{ K W}^{-1}.$$

TABLE 5.3. SUMMARY OF COMPONENT HEAT LOSSES

Outside air temperature 11 °C
Inside air temperature 25 °C

Component	Structure	Area (m²)	Overall heat Transfer coefficient U (W m⁻² K⁻¹)	Thermal resistance R (m² K W⁻¹)	Heat loss Q (kW)	Percentage of total loss
Base	500 mm thick solid concrete	100	0.42	2.38	0.17	2.5
Walls	Unventilated air-filled cavity built 100/50/100 mm	280	0.48	2.08	1.88	28.2
Glazing	6 mm thick single panes	120	2.00	0.49	3.36	50.4
Roof	Flat 150 mm air cavity between 10 mm tiles with 3 mm felt and 10 mm ceiling board	100	0.89	1.12	1.25	18.9
Effective total values	—	600	0.79	1.26	6.66	100

5.9   INFILTRATION AND VENTILATION

Air movements by infiltration through an insulated structure or by controlled ventilation imposes extra loads on air-conditioning systems because outside air must be heated or cooled to the condition prevailing in the internal environment. Infiltration losses from a building occur when air passes through gaps in and around doors and windows. The rate of air flow is approximately proportional to the square of the pressure difference $\sum p$ (N m⁻²) acting across the component involved [1], i.e.

$$\dot{V} \propto A(\Delta p)^n, \tag{5.16}$$

where $\dot{V}$ is the rate of air flow (m³ s⁻¹) and A is the area of the aperture (m²). The constant of proportionality ($\simeq 0.5$–0.7) and the value of the index n ($\simeq 0.5$) depends upon the characteristics of the component involved [19]. The static pressure on the windward side of a building ranges from about 0.5 to 0.8 times the velocity pressure ($= \rho u^2/2$) of the undisturbed free flow of wind. On the leeward side, an external "negative pressure" of approximately 0.3–0.4 times the free flow velocity pressure is induced. The local pressure varies in a complicated fashion over the sides of the building parallel to the wind direction, depending partly upon the velocity gradients and the aerodynamic characteristics of the surfaces, and being affected by the height of the building and the sizes, shapes, and proximities of neighbouring structures. Mean pressure differences across buildings of different heights for a free wind speed of 9 m s⁻¹ are tabulated in Table 5.4 [19].

Natural infiltration in most existing buildings is often far greater than the minimum ventilation requirements and so unnecessary burdens to conditioning equipment. Any integrated design procedure for energy-saving arrangements cannot allow the introduction of an unknown quantity. Infiltration is essentially a random occurrence depending upon uncontrollable factors. For a controlled situation, infiltration must be eliminated by draught-proofing

TABLE 5.4. PRESSURE DIFFERENCES DUE TO WIND EFFECTS [19]

Building height (m)	Mean pressure difference across a building (N m^{-2})		
	Open country (9 m s^{-1})	Suburban (5.5 m s^{-1})	City centre (3 m s^{-1})
10	58	21	6
20	70	31	11
30	78	38	15
40	85	44	21
50	90	49	23
60	95	55	26
70	100	59	31
80	104	63	34

and associated isolating techniques. Outside air introduction must be regulated and restricted to relate to ventilating requirements. Recommended ventilation rates vary according to the rates of oxygen consumption within the building and are thus dependent upon rates of metabolism, combustion in open fires, cigarette smoking and cooking, and odour and contaminant production. Not less than 21 m^3 h^{-1} of fresh air per person is advised [19] to prevent vitiation and palpable body odour. Some recommendations for minimum requirements are listed in Table 5.5.

Thermal load due to ventilation. The thermal requirements of the fresh air introduced for ventilation purposes may be calculated from

$$\dot{Q}_a = \dot{m}\Delta H, \qquad (5.17)$$

where $\dot{m}$ (kg s^{-1}) is the air mass flow and ΔH is the difference in specific enthalpy (J kg^{-1}) between the outside environmental air and the inside air condition. If no change in specific humidity is achieved the air undergoes sensible heating or cooling alone, i.e.

$$\dot{Q}_a = \dot{m}c_p\Delta T, \qquad (5.18)$$

where c_p is the mean specific heat (J kg^{-1} K^{-1}) of the air and ΔT is the difference between the dry-bulb temperatures outside and inside.

Equation (5.24) may be incorporated into Fig. 5.4 by defining an effective transfer function representative of ventilating needs, i.e.

$$\dot{Q}_a = \dot{m}c_p(T_i - T_o) \equiv U_a A(T_i - T_o)$$

TABLE 5.5. RECOMMENDED AIR CHANGES [19]

Occupancy known		Occupancy unknown	
Type	Air changes (m^3/person/s)	Type	Air changes/hour
Homes	0.012	Offices	3–8
Schools, theatres	0.014	Engine rooms	4
Factories, shops	0.016–0.028	Garages	5
Hospitals	0.019–0.047	Baths	5–8
		Lavatories	5–10
		Restaurants	5–10
		Cinemas, theatres	5–10
		Kitchens	10–40

and so

$$U_a = \frac{\dot{m}c_p}{A}, \tag{5.19}$$

where A (m²) is the total surface area of the structure. Alternatively an effective resistance may be introduced,

$$R_a = \frac{A}{\dot{m}c_p} \tag{5.20}$$

For the system under consideration,

$$A = 600 \text{ m}^2, \qquad \dot{m} \simeq 0.012 \text{ m}^3/\text{person/s}$$

$$= 0.036 \text{ m}^3 \text{ s}^{-1} \text{ for three persons,}$$

and $c_p = 1004.4 \text{ J kg}^{-1} \text{ K}^{-1},$

and the mean air density $= 1.182 \text{ kg m}^{-2}.$

Thus $U_a = 0.07 \text{ W m}^{-2} \text{ K}^{-1}$

and $R_a = 14.04 \text{ m}^2 \text{ K W}^{-1}$

and so $\dot{Q}_a = 588 \text{ W} = 0.588 \text{ kW}.$

5.10 CONCLUSIONS

The foregoing consideration is an approximate analysis of the thermal demands of buildings. The results obtained include the transmission and ventilating losses from the structure described under a preset temperature excess representing an annual mean situation. Effects of room air circulation and stratification, heat losses through apertures, transient thermal flows, localised room air temperature variations, intermittent demands, and heat gains have not been considered. These are discussed in a later chapter. The study has, however, illustrated methods of analysis and produced data which enables order of magnitude component heat loads to be appreciated. Effects of insulating techniques are quantified in Chapter 8. The calculations have shown that 50% of the total energy lost by transmission from the building considered travels through its windows (assumed 30% glazed), 28% through the walls, 19% through the flat roof, and the remaining 3% is transmitted through the base (see Table 5.3). A ventilating system requires a further 0.6 kW (Table 5.6).

TABLE 5.6. HEAT LOSSES FROM BUILDING COMPONENTS
(REPRESENTATIVE FIGURES)

Component	Average annual heat loss	
	kW	% of total
Base	0.17	2.3
Walls	1.88	26.0
Glazing	3.36	46.4
Roof	1.25	17.3
Ventilation	0.58	8.0
Totals	7.24	100.0

APPENDIX
HEAT TRANSFER THROUGH A PITCHED ROOF

A pitched roof (Fig. A.5.1) contains essentially three thermal resistances in series: the thermal resistance of the ceiling, that of the air space in the attic, and the resistance of the sloping roof.

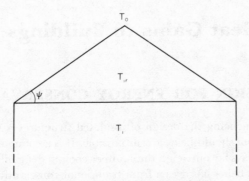

Fig. A.5.1. Sketch of a pitched roof.

Under steady-state conditions:

Heat flow through the ceiling to the roof space	heat flow from the roof space = to the environment	the overall heat flow from = the living space to the environment through the composite roof

$$U_{cg}A_{cg}(T_i - T_{irf}) \quad = \quad U_{rf}\frac{A_{cg}}{\cos\psi}(T_{irf} - T_o) \quad = \quad UA_{cg}(T_i - T_o) \tag{A5.1}$$

where U_{cg} and U_{rf} are the thermal conductances of the ceiling and roof respectively (W m^{-2} K^{-1}), A_{cg} is the area of the ceiling (m^2), T_i and T_o are the inside and outside environmental temperatures (°C), T_{irf} is the temperature (assumed uniform) of the air in the roof space (°C), and U is the effective overall conductance across the structure (W m^{-2} K^{-1}).

Solving eqn. (A5.1) for the overall heat transfer coefficient U,

$$U = \frac{U_{cg}U_{rf}}{U_{rf} + U_{cg}\cos\psi}. \tag{A5.2}$$

Chapter 6

Heat Gains to Buildings

6.1 DESIGN FOR ENERGY CONSERVATION

A primary criterion influencing the design of insulated structures is that the total energy requirements of the system should be as low as possible. It is a common misconception that winter heating loads always far outweigh the summer cooling demands of buildings sited in the British climate. This belief has arisen from traditional conservative practices in design and construction engendered by a long, unbroken period of unrestricted cheap energy supply. Alternatively, design considerations founded upon predicted maximum heat losses could cause unnecessarily high levels of insulation to be applied. The indiscriminate application of thermal insulation can lead to an increase in annual energy consumption by introducing a need to cool the structure during warmer seasons. The adoption of air-conditioning plant capacities, based upon predicted maximum heat gains, results in under-used capital equipment and the lower efficiencies associated with off-design operation. A complete analysis of expected net heat flows during a period representative of the yearly climatic cycle is desirable. Thermal energy storage facilities should be integral to any building service arrangement.

It is possible to envisage buildings with walls which inhibit transmission losses whilst encouraging solar radiation gains. Excess heat can be redirected and stored to be released during colder spells. No internal heat generation or removal would be needed to maintain a thermally comfortable internal environment throughout the year. The building would thus be self-sufficient in the energy required for heating purposes. The design problems to be overcome are far less formidable in temperate zones than those encountered in less-favourable climatic conditions where either excessive heating or refrigerating is stipulated.

In order to appraise these possibilities the variation of solar influx over an annual cycle at a location in the United Kingdom will be examined, and total annual solar and sundry heat gains associated with the building previously described in Chapter 5 will be quantified. Thus the net annual thermal loading will be estimated.

6.2 SOLAR DATA

6.2.1 Summary of pertinent relationships

Declination, d'. This varies from $\pm 23\frac{1}{2}°$ on solstices (i.e. with the sun over the equator) to $0°$ at equinoxes (cf. section 4.5).

Altitude α′. At noon,

$$\alpha' = 90 - (l' - d'). \tag{4.2}$$

At other times,

$$\sin \alpha' = \sin d' \sin l' + \cos d' \cos l' \cos h', \tag{4.3}$$

where the hour angle

$$h' = 15t° \text{ (sun-time)}. \tag{4.4}$$

Figure 6.1 shows the variation in the altitude angle for the latitude 51.7°N.

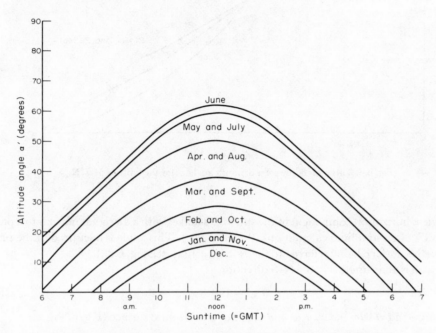

Fig. 6.1. Variation of the solar altitude angle α′ for the latitude 51.7°N.

Azimuth z′. At noon,

$$z' = 0.$$

At other times,

$$\tan z' = \frac{\sin h'}{\sin l' \cos h' - \cos l' \tan d'}. \tag{4.5}$$

Figure 6.2 shows how the azimuth angle varies at the latitude 51.7°N.

Direct solar radiation I_δ

$$I_\delta = K_1 \exp(- K_2/\sin \alpha'). \tag{4.6}$$

Direct solar radiation can be considered as consisting of straight parallel rays when it reaches the earth. When this flux encounters a surface the amount of radiation intercepted can be calculated from simple geometrical laws.

Solar wall azimuth angle n′. The use of the solar wall azimuth angle n', which is defined as [19]

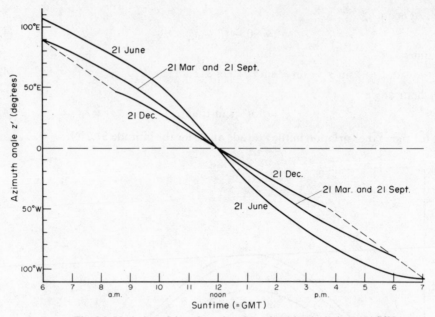

Fig. 6.2. Variation of the solar azimuth angle z' for the latitude 51.7°N.

"the angle a horizontal component of the sun's rays makes with a direction normal to a particular wall", often simplifies calculations ($0 \leq n' \leq 90°$). This angle is measured in the plane of the earth's surface being referred to the N–S direction (see Fig. 6.3a).

If the wall is at an angle ψ to the N–S direction,

$$n' = z' - (90 - \psi). \tag{6.1}$$

The intensity of direct solar radiation on a surface. For a horizontal surface (Fig. 6.3b),

$$I_H = I \sin \alpha'. \tag{6.2}$$

For a vertical surface (Fig. 6.3c),

$$I_V = I \cos \alpha' \cos n'. \tag{6.3}$$

For a surface tilted at θ to the horizontal (Fig. 6.3d),

$$I_\theta = I \sin \alpha' \cos \theta \pm I \cos \alpha' \cos n' \sin \theta \tag{6.4}$$

or

$$I_\theta = I \cos i', \tag{6.5}$$

where i' is the angle of incidence.

Figure (6.4) indicates the variations of direct solar radiation with altitude and azimuth angles for various surface inclinations.

Scattered radiation I_{sc} [24]. The relationship for scattered sky radiation (eqn. (4.7)) may be extended to include the radiation reflected from the ground and surrounding surfaces which is received by a wall inclined at an angle θ to the horizontal.

$$I_{sc} = K_3 I_\delta F_{sc} + \mathscr{R}_{gd} F_\delta (K_3 + \sin \theta) F_{gd} \tag{6.6}$$

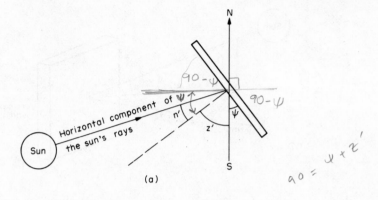

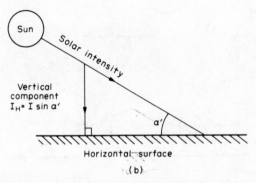

Fig. 6.3. Solar radiation striking surfaces of various inclinations: (a) definition of the solar wall azimuth angle; (b) incidence on a horizontal surface; (c) incidence on a vertical surface; (d) incidence on a pitched roof. $a \equiv I \sin \alpha'$, $b \equiv I \sin \alpha' \cos \theta$, $c \equiv I \cos \alpha' \cos n'$, $d \equiv I \cos \alpha' \cos n' \sin \theta$.

for which values for the constant K_3 are listed in Table 4.3. F_{sc} and F_{gd} are angle factors for radiation reflected from the surrounding surfaces and the ground respectively, and $\mathcal{R}_{gd}$ is the reflectivity of the ground ($= 0.2$ for grass, $= 0.7$ for white stones). For a surface seeing only ground and sky [45],

$$F_{gd} = 0.5 \quad (1 - \cos \theta), \tag{6.7}$$

$$F_{sc} = (1 - F_{gd}). \tag{6.8}$$

Variations in the intensity of diffuse radiation with altitude angle for vertical and horizontal surfaces are plotted in Fig. 6.5.

6.2.2 **Total solar radiation received by a surface**

The total rate of interception of solar radiation consists of the sum of the direct sky-scattered and ground-reflected components:

$$I_{tot} = I_\delta + I_{sc} + I_{gd}. \tag{6.9}$$

Total solar intensities on vertical N, S, E, W, NE, NW, SE, SW facing and horizontal surfaces for the latitude 51.7°N are given in Fig. 6.6a–e.

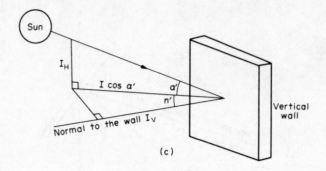

(c)

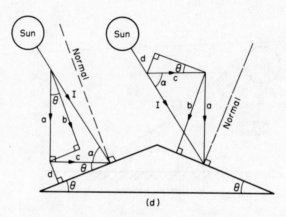

(d)

Fig. 6.3 (c) and (d)

TABLE 6.1. MONTHLY MEAN INSOLATION (51.7°N) (W m^{-2})

	Monthly mean values of insolation on surfaces orientated as indicated (W m^{-2})								
Month	Horizontal	N	S	E	W	NW	NE	SE	SW
January	50	10	130	35	35	10	10	95	95
February	95	20	180	70	70	25	25	135	135
March	165	30	200	110	110	50	50	170	170
April	245	50	185	150	150	90	90	190	190
May	305	75	165	180	180	125	125	190	190
June	330	10	155	190	190	140	140	185	185
July	305	75	165	180	180	125	125	190	190
August	245	50	185	150	150	90	90	190	190
September	165	30	200	110	110	50	50	170	170
October	95	20	180	70	70	25	25	135	135
November	50	10	130	35	35	10	10	95	95
December	35	10	105	25	25	10	10	75	75

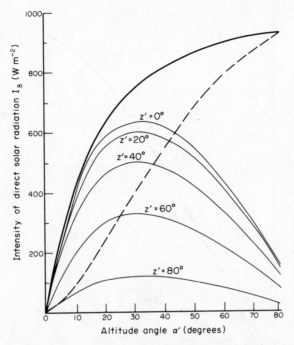

Fig. 6.4. Variation of the intensity of direct solar radiation (W m^{-2}) with altitude and azimuth angles for various surface inclinations (clear sky, 0–300 m above sea-level). ——— on vertical walls, ——— normal to the sun; ———— on a horizontal surface.

6.2.3 **Application of solar data to buildings**

It is generally recommended that a solar energy collecting arrangement should be situated on the south-facing wall of a building. If, however, the building as a whole is intended to be a solar collecting unit, net heat balance calculations must consider heat flows through all four vertical walls and the roof. Whilst a south-facing wall often intercepts the maximum solar radiation, the north-facing in general receives the least solar heat and also often suffers large transmission losses. A building with a south-facing wall must also contain a north-facing wall of the same area. Figure 6.7a–e demonstrates net effects of wall combinations for a N–S oriented building. Table 6.1 lists monthly mean insolation values and monthly and yearly totals have been derived (Table 6.2). It may be seen that, in the absence of cloud cover, a cubical structure orientated NS–EW intercepts annually an amount of solar radiation equivalent to ~ 120 W m^{-2} continuously. Table 5.6 showed that the same shaped structure loses on annual average ~ 12 W m^{-2} by thermal transmission. Thus if only 10% of the solar radiation incident on the building (in the absence of cloud cover) could be collected and stored, the overall heating requirements to offset annual transmission losses could be provided completely by the sun.

6.2.4 **Optimising building shape with respect to solar interception**

The total annual insolation received by the building *abc* (Fig. 6.8) is given by (from Table 6.2)

$$I = 5492\ ac + 6437\ ab + 6872\ bc\ \text{MJ m}^{-2}. \tag{6.10}$$

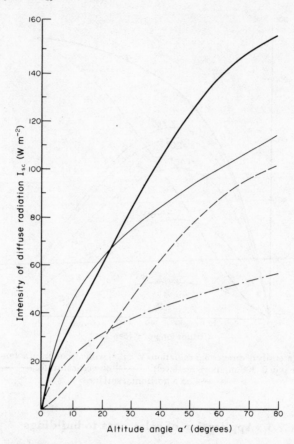

Fig. 6.5. Variation of the intensity of diffuse solar radiation (W m^{-2}) for vertical and horizontal surfaces as a function of altitude angle. ——— total of diffuse sky radiation plus radiation reflected from the ground on to a vertical wall; ——— on a horizontal surface; ——— radiation reflected from the ground and received at the surface of a vertical wall (ground reflectance factor = 0.2); —.— on a vertical wall.

TABLE 6.2. TOTAL MEAN INSOLATION (51.7°N) (MJ m^{-2})

Month	Horizontal	N	S	E	W	Monthly average total for a cubical structure
December	94	27	281	67	67	107
January and November	264	53	685	185	185	274
February and October	484	102	918	357	357	444
March and September	870	158	1054	580	580	648
April and August	1291	263	975	791	791	822
May and July	1634	402	884	964	964	970
June	855	233	402	492	492	495
Yearly totals	5492	1238	5199	3436	3436	3760

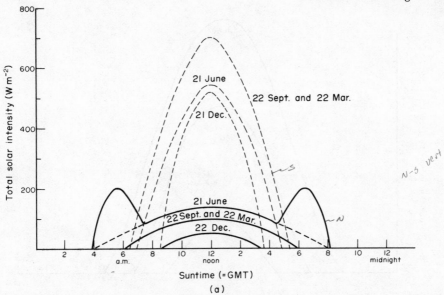

Fig. 6.6. Total solar radiant intensities (direct plus diffuse) striking vertical and horizontal surfaces at the latitude 51.7°N: (a) for north- (———) and south- (———) facing walls; (b) for east- and west-facing walls; (c) for a horizontal surface; (d) for north-east- and north-west-facing walls; (e) for south-east- and south-west-facing walls.

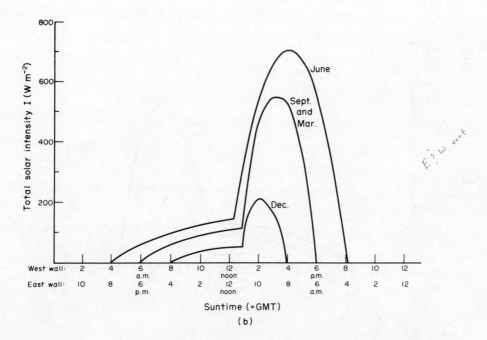

Fig. 6.6b

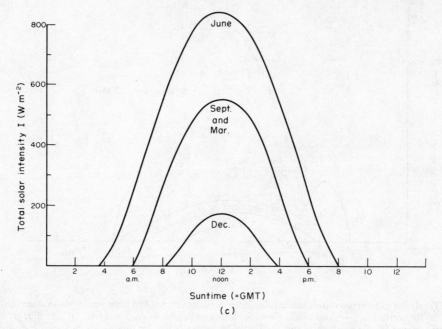

Fig. 6.6c

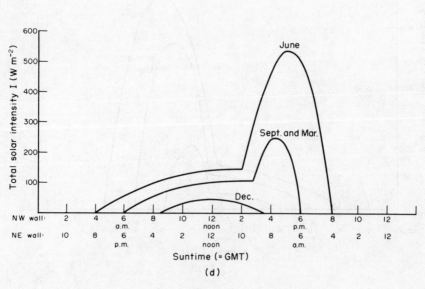

Fig. 6.6d

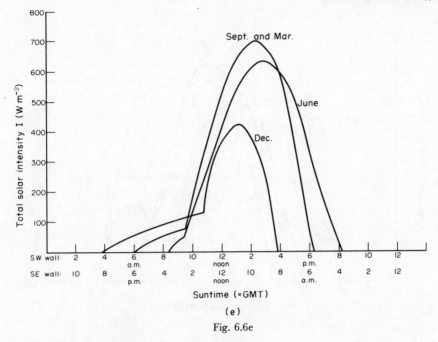

Fig. 6.6e

It is interesting to observe that the total solar radiation intercepted by the south plus north walls is less than that received by the east and west walls. In designing for solar energy pick-up, it appears therefore that the east and west faces should be constructed to be longer than the north and south faces. Equation (6.10) has been plotted for 5 m and 10 m high buildings each of volume 1000 m^3 (Fig. 6.9). If the smallest practical plan dimension is 5 m:

For maximum solar interception:

For the 5 m high building:

$a = 5$ m; $b = 5$ m; $c = 40$ m.

$I = 2.75 \times 10^6$ MJ over 650 m^2 of exposed wall area.

$I/A = 4230$ MJ m^{-2}.

For the 10 m high building:

$a = 5$ m; $b = 10$ m; $c = 20$ m.

$I = 2.25 \times 10^6$ MJ over 600 m^2 of exposed wall area.

$I/A = 3750$ MJ m^{-2}.

In each case the longest side of the building faces west. The lower profile structure is evidently preferable.

For minimum solar interception:

For the 5 m high building:

$a = 14.6$ m; $b = 5$m; $c = 13.7$ m.

$I = 2.04 \times 10^6$ MJ over 483 m^2 of exposed wall area.

$I/A = 4220$ MJ m^{-2}.

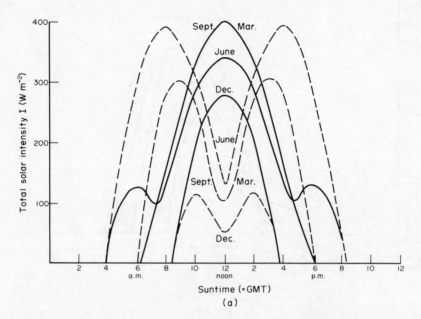

Fig. 6.7. Total solar radiant intensities (direct plus diffuse) striking the external surfaces of buildings at the latitude 51.7°N: (a) north- plus south-facing walls (————), east- plus west-facing walls (— — —); (b) north- plus south- plus east- plus west-facing walls; (c) as (b) plus the influx on a flat roof; (d) annual variations of mean daily total solar influxes on single surfaces; (e) annual variations of mean daily total solar influxes on combinations of surfaces.

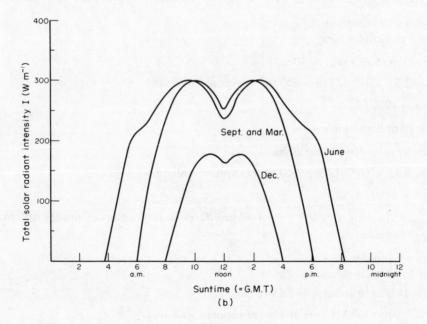

Fig. 6.7b

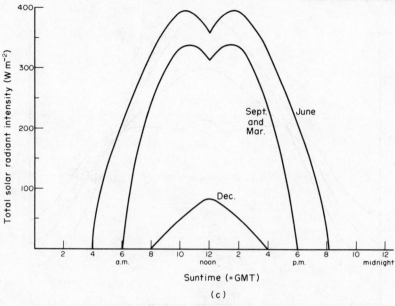

Suntime (=GMT)

(c)

Fig. 6.7c

For the 10 m high building:

$a = 10.3$ m; $b = 10$ m; $c = 9.7$ m.
$I = 1.88 \times 10^6$ MJ over 500 m² of exposed wall area.
$I/A = 3760$ MJ m⁻².

Given overall heat transfer coefficients U for the walls, transmission heat losses depend upon the area of exposed walling. Hence the ratio I/A indicates relative net gains. In order to quantify net heat transfers it is necessary to specify more fully the construction details. The characteristics of structural components with respect to solar radiant energy are first considered.

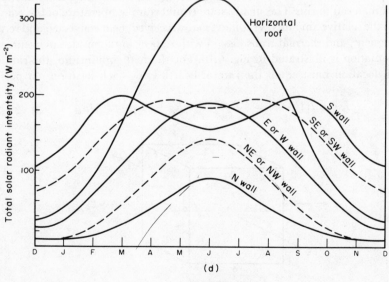

(d)

Fig. 6.7d

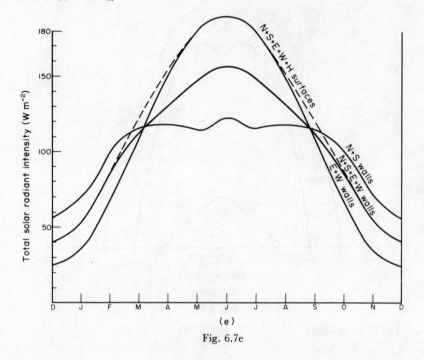

(e)

Fig. 6.7e

6.3 NET HEAT TRANSFERS THROUGH BUILDING COMPONENTS

6.3.1 Wall and solid roofs

A vertical wall or a roof receives solar radiation at its outer surface. Some of this energy is reflected away from the surface and the remainder is absorbed causing the temperature of its constituent material to rise. The steady-state equilibrium temperature of the component depends upon the relative amounts of radiant energy received, heat convected and reradiated to the environment, and thermal transmission with respect to the inside environment. The steady-state situation is illustrated in Fig. 6.10. For thermal equilibrium, the rate of heat entering each location must equal the rate of heat leaving each location. At position 2, therefore,

$$I\mathscr{A} = I(1 - \mathscr{R}) = (T_2 - T_3)/R_o + (T_2 - T_4)/R_b, \tag{6.11}$$

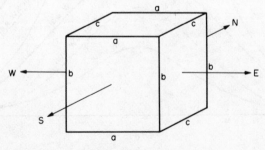

Fig. 6.8. Sketch of the building shape considered for shape optimisation.

where $\mathscr{A}$ is the surface absorptivity to solar radiation and $\mathscr{R}$ is the surface reflectivity to solar radiation. Similarly, at position 4,

$$(T_2 - T_4)/R_b = (T_4 - T_5)/R_i = \dot{q}_i, \qquad (6.12)$$

where $\dot{q}_i$ is the mean rate of internal heat generation or removal to maintain the internal environment at a constant temperature. In design for minimum energy expenditure, $\dot{q}_i$ should include the internal sundry heat gains (cf. section 6.4) arising from lighting, metabolisms, and power dissipation from electrical equipment.

The sol-air temperature. The IHVE [19] have argued that absorbed solar radiation has the same effect as a rise in outside temperature and so have recommended the use of the sol-air temperature concept to simplify calculations. The sol-air temperature is defined as the outside air temperature which, in the absence of solar radiation, would give the same temperature distribution and rate of heat transfer through a wall as exists with the actual outdoor temperature and the incident solar radiation. The sol-air temperature at a particular time can be calculated from

$$T_{sa} = T_o + R_o(\mathscr{A}I - \varepsilon I_l), \qquad (6.12a)$$

where I is the intensity of direct plus diffuse solar radiation on the outer surface (W m^{-2}); I_l is the intensity of long-wave radiation from a black surface at the temperature of the environmental air ($= 100$ W m^{-2} for radiation from a horizontal roof to a cloudless sky [19]; = zero for a vertical wall because it is assumed that [19] radiative gain from the ground

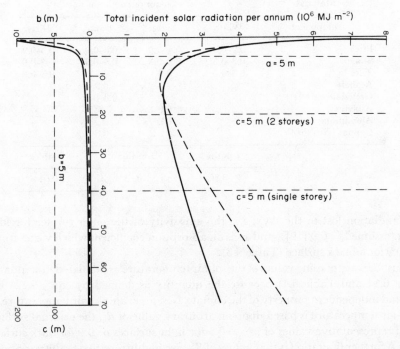

Fig. 6.9. Optimising building shape for maximum and minimum solar gains: ——— single-storey building; ——— two-storey building.

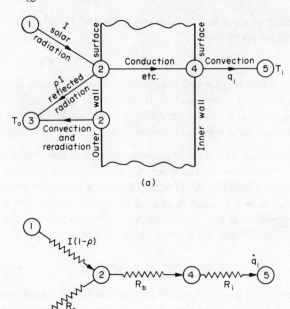

Fig. 6.10. Steady-state heat flows at a wall: (a) physical situation; (b) equivalent thermal network.

TABLE 6.3. ABSORPTION COEFFICIENTS $\mathscr{A}$ [19]

Material or component	Colour or condition			
	White	Light	Dark	Dirty
Brick	0.2–0.5	0.4–0.5	0.6–0.9	0.5–0.9
Stone	0.3–0.5	0.3–0.5	0.5–0.6	0.5–0.9
Tiles	0.3–0.5	0.4	0.8	0.5–0.9
Asphalt			0.9	
Grey slate			0.8–0.9	
Asbestos		0.6		
Aluminium		0.2		
Copper (tarnished)			0.6	

Water	1 m thick	2 m thick	3 m thick
	0.56	0.61	0.64

balances radiation lost to the sky); ε is the emissivity of the outer surface for long-wave radiation (assumed $= 0.9$ [19]), and $\mathscr{A}$ is an absorption coefficient which varies from 0.5 for brick to 0.9 for a black surface (Table 6.3).

One must be wary of using values of the sol-air temperature as tabulated in standard guides [19] (Fig. 6.11 and Table 6.4) because the quantity as defined by eqn. (5.12) is not an intrinsic and independent property of the climate as is, for example, air temperature. A value of the sol-air temperature is based upon an arbitrary value of R_o, the external surface resistance, and representative values of $\mathscr{A}$ ($=0.5$ for light surfaces or 0.9 for dark surfaces) and ε ($=0.9$). A further limitation is that values of T_{sa} are highly sensitive to the variations in the external surface resistance R_o which itself is strongly affected by wind speeds.

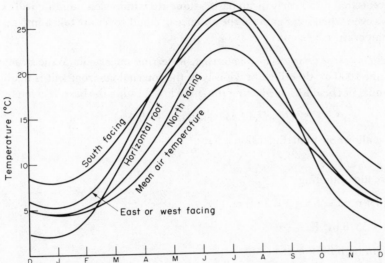

Fig. 6.11. The annual variation in the mean daily sol-air temperatures for "dark" surfaces facing in the directions indicated (latitude 51.7°N in the United Kingdom).

For a thermally black surface,

$$T_{sa} = T_o + R_o(I - I_l),\qquad(6.13)$$

and for an ideal thermal rectifier (cf. Chapter 11),

$$T_{sa} = T_o + R_oI,\qquad(6.14)$$

where $R_o \rightarrow \infty$. In theory, the sol-air temperature can achieve very high values providing that solar heat gains are enhanced and heat losses are inhibited.

Having obtained a value for the sol-air temperature, the steady-state heat transfer through the wall is calculated from

$$Q = UA(T_i \sim T_{sa}).\qquad(6.15)$$

TABLE 6.4. MEAN REPRESENTATIVE SOL-AIR TEMPERATURES T_{sa} [19]

Month	Mean sol-air temperature (°C)					Mean air temperature (°C)
	Horizontal roof	North wall	South wall	East wall	West wall	
December	2.5	5.5	8.5	6.0	6.0	5.0
January	2.0	4.5	8.0	5.0	5.0	4.3
February	3.6	5.5	9.6	7.0	7.0	5.0
March	8.0	7.5	12.5	10.5	10.5	6.5
April	14.0	11.0	16.5	16.5	16.5	9.0
May	20.5	16.5	20.5	21.5	21.5	13.5
June	25.0	21.0	23.5	26.0	27.5	16.5
July	27.0	22.5	26.0	27.6	27.5	19.0
August	22.0	19.0	26.0	24.0	24.0	17.0
September	15.5	15.0	21.5	19.0	19.0	13.0
October	7.5	10.5	15.5	12.5	12.5	10.0
November	4.0	7.0	11.5	7.5	7.5	6.5
Annual mean values (°C)	12.64	12.6	16.3	15.3	15.3	11.0

Table 6.7 reveals that on a yearly mean basis a standard windowless building will experience a net heat loss even when solar gains in the absence of cloud cover are taken into account via the sol-air temperature concept.

Optimising building shape using sol-air temperatures. Referring once again to the structure *abc* of internal volume 1000 m³ described previously, if the internal environment is maintained at 25°C, the annual net heat loss, neglecting the small loss through the base, is given by

$$Q = U(12.4ac + 21.7ab + 19.4bc). \tag{6.16}$$

If the smallest allowable plan dimension is 5 m:

For minimum net heat loss:

For the 5 m high building:

$a = 13.4$ m; $b = 5$ m; $c = 14.9$ m.

$Q/U = 5378$ m² K.

For the 10 m high building:

$a = 9.5$ m; $b = 10$ m; $c = 10.5$ m.

$Q/U = 5338$ m² K.

Thus, when both heat losses and solar gains are considered, the optimum floor shape for minimum net heat loss approaches the square configuration necessary for minimum rates of transmission loss when the overall heat transfer coefficient U is uniform over the vertical surfaces of the structure.

6.3.2 Glazing

When solar radiation encounters a window, part of its energy is transmitted directly to the interior, part is absorbed by the glass to be subsequently convected and reradiated from its surface, and the remainder is reflected. The relative amount of transmission, absorption, and

TABLE 6.5. CHARACTERISTICS OF GLAZING [19]

	Type		Radiation transmitted and retransmitted, Γ				
			Direct radiation for given angle of incidence (degrees)				Diffuse radiation
			0	30	60	85	
Single	Clear 4 mm		0.84	0.84	0.77	0.22	0.78
	Clear 6 mm		0.80	0.79	0.72	0.20	0.73
	Heat absorbing 6 mm		0.50	0.49	0.36	0.13	0.44
	Heat reflecting		0.27	0.27	0.26	0.11	0.25
Double	Clear 4 mm	Clear 4 mm	0.75	0.73	0.65	0.12	0.67
	Clear 6 mm	Clear 6 mm	0.69	0.67	0.59	0.11	0.60
	Heat absorbing	Clear 4 mm	0.45	0.44	0.37	0.10	0.37
	Heat absorbing	Clear 6 mm	0.37	0.35	0.29	0.08	0.30
	Heat reflecting	Clear	0.33	0.28	0.25	0.07	0.28
	Outer pane	Inner pane					

reflection depends upon the angle of incidence of the incoming radiation and the values of transmissivity, absorptivity, and reflectivity, $\mathscr{T}$, $\mathscr{A}$, and $\mathscr{R}$, which are applicable for the glass with respect to short-wave solar radiation. The figures for transmittance Γ listed in Table 6.5 each represent the total fraction of the incident solar energy which reaches the interior of the building by direct transmittance *and by retransmission*. Calculations which compute total heat gain by integrating instantaneous solar gains for given altitude and azimuth angles are tedious. The IHVE [19] have therefore recommended the use of mean solar gain factors, S, for various types of glazing and shading (Table 6.6).

TABLE 6.6. MEAN SOLAR GAIN FACTORS FOR UK APPLICATION [19]

Type of glazing and protection	Solar gain factors S	
	Single	Double
Clear glass with no shading	0.76	0.64
Heat-absorbing glass with no shading	0.45	0.31
Heat-reflecting glass with no shading	0.26	0.25
Clear glass with an internal white venetian blind	0.46	0.46
Clear glass with an internal white cotton curtain	0.41	0.40
Clear glass with mid-pane white venetian blind	—	0.28
Clear glass with an external canvas roller blind	0.14	0.11
Clear glass with an external white louvred sun-breaker	0.14	0.11

Any object placed between the sun's rays and the receiving surface reflects some radiation, absorbs some radiation, and transmits the rest. Whilst diffuse sky radiation is received by the whole area, shaded or unshaded, fractions of incident direct solar radiation cut-off by lintels, canopies, window recesses, or adjacent buildings must be calculated using the laws of geometrical optics [19, 24]. Internal shading devices (curtains, blinds, etc.) reduce instantaneous solar gains, but, because this heat is later retransmitted by convection and reradiation, these cannot be used to regulate the overall internal heat balance unless some means of direct control is provided. External arrangements (canopies, louvred, or slatted blinds, etc.) are more effective because the secondary heat transmission then takes place to the external environment.

6.4 NET HEAT BALANCE OF A BUILDING

Transmission heat losses for a building in the absence of solar radiation were estimated in Chapter 5. The structure defined then will now be further considered to ascertain its total annual thermal loading.

6.4.1 Construction details

In addition to the information given in Table 5.2 it is necessary to specify the positions and details of the glazing. A proportion of glazing of 30% has already been assumed. It is now further assumed that this is equally distributed on all four vertical walls of the building (Table 6.7).

TABLE 6.7. MEAN NET HEAT LOSSES OVER AN ANNUAL PERIOD

Component	Area (m^2)	Overall heat transfer coefficient U $(W\,m^{-2}\,K^{-1})$	Solar gain factor S	Mean annual sol-air temperature $\bar{T}_{sa}$ $(°C)$	Annual mean insolation $\bar{I}$ $(W\,m^{-2})$	Mean net heat gain (W)
Base	100	0.42	—	—	—	−168
South wall	70	0.48	—	16.3	165	−292
North wall	70	0.48	—	12.0	39	−437
East wall	70	0.48	—	15.3	109	−326
West wall	70	0.48	—	15.3	109	−326
South glazing	30	2.00	0.76	—	165	+2922
North glazing	30	2.00	0.76	—	39	+50
East glazing	30	2.00	0.76	—	109	+1645
West glazing	30	2.00	0.76	—	109	+1645
Roof	100	0.89	—	12.6	175	−1104
Total						+3901

6.4.2 **Mean net heat gains from the external environment**

If the interior is maintained at 25°C, the mean net heat gain $\dot{Q}$ over an annual period in the absence of cloud can be calculated from:

For the base:

$$\dot{Q}_b = - UA(T_i - \bar{T}_o\,(+10°C)). \qquad (6.17)$$

For the walls and roof:

$$\dot{Q}_{w,rf} - UA(T_i - \bar{T}_{sa}). \qquad (6.18)$$

For the glazing (neglecting external shading):

$$\dot{Q}_g = SA\bar{I} - UA(T_i - \bar{T}_o). \qquad (6.19)$$

Table 6.7 shows that, on an annual mean basis, the building considered experiences a net heat gain equivalent to a continuous power rating of 3.9 kW.† Reductions in this figure resulting from intermittent local cloud cover and other shading effects must be estimated using historical climatic, geographic, and topological data for a particular location.

Over the annual climatic cycle a building is seldom in thermal equilibrium with its environment. In summer, solar heat must be excluded or else the building interior must be cooled—at least by opening windows, whereas in winter, when transmission losses overwhelm solar gains, substantial internal heating is required. Internal environmental heating and cooling systems must be designed to be able to cope with the extremes of climate (Table 6.8).‡ The desirability of long-term thermal storage facilities for excess solar heat is clear. Possibilities of solar collection and storage are investigated in Chapter 11.

---

†The effects of size, shape, and position of windows are further discussed in Chapter 8.
‡Time lags and decrement factors have been neglected in assessing the summer high design situation shown in Table 6.8. These effects are treated in Chapter 7. The winter loss data has been calculated assuming an outside environmental temperature of −5 °C. The overall heat transfer coefficients adopted for the walls, glazing, and roof include modifications for the effects of a high wind speed and radiative heat transfers (cf. sections 2.6.3 and 5.6). Hence, in all cases, the relationship

$$\dot{Q} = UA\,(T_i - T_o) = 30UA\ \text{W}$$

was used to calculate the values.

TABLE 6.8. ANNUAL EXTREME CONDITIONS

Component	Summer high design situation (23 July at 1500 h)[a]			Winter low design situation ($T_o = -5°C$ at night)
	Sol-air temperature T_{sa} (°C)	Solar intensity I (W m^{-2})	Heat gain (W)	Heat gain (W)
Base	—	—	0	−840
South wall	41.2	345	+544	−1008
North wall	30.2	105	+175	−1008
East wall	30.2	105	+175	−1008
West wall	53.7	630	+964	−1008
South glazing	—	345	+7880	−1800
North glazing	—	105	+2330	−1810
East glazing	—	105	+2330	−1800
West glazing	—	630	+14300	−1800
Roof	44.3	630	+1718	−2670
Totals			+30416	−13902

[a]Outside environmental air temperature = 25.5°C. Values taken from reference [19].

6.4.3 Sundry heat gains

The heat balance of the building is completed by so-called "bonus" heat gains arising from electrical equipment and metabolic and process heat dissipation. An estimate of total sundry gains may be obtained by monitoring electricity and fuel consumption. All electricity converted to work, sound, or light appears eventually as a thermal load on the interior. The heat given out by equipment fuelled by alternative forms of energy depends upon the conversion efficiencies of the various devices (cf. Chapter 8).

Electrical lighting. Much of the energy supplied to electric lights appears immediately as heat, but even the proportion initially dissipated as short-wave light finally becomes longer-wave thermal energy after multiple re-reflections and reactions with the surfaces inside the building (Table 6.9).

Occupants. Human beings reject energy as both sensible and latent heat. The total amount of heat and the proportions involved depend upon the degree of metabolic activity which is

TABLE 6.9. HEAT DISSIPATION FROM LAMPS (*approximate values*)

Illumination on the working plane (lux)	Wattage per m² of floor area (including power for control gear)			
	Filament lamps		80 W white fluorescent[a]	
	With reflector	With diffuser	In diffusing fitting	In louvred ceiling panel
150	19–28	28–36	8	8–11
200	28–36	36–50	11	11
300	38–55	50–69	11–16	14–19
500	66–88	—	22–28	22–33
1000	—	—	36–55	44–66

[a]An 80 W fitting needs 100 W of power supplied to it; the surplus of 20 W is liberated directly from the control gear as heat.

related to the rate of working and the environmental air condition (from $\sim 25\%$ latent heat for a basal metabolic rate in an environment at effectively 20°C to over 75% latent heat during heavy work (see Table 3.1). Densities of occupation are usually assumed [19] when exact figures are not known, i.e.

10 m^2 per person for an office block

20 m^2 per person in executive offices

2 m^2 per person in restaurants

0.5 m^2 per person in cinemas and theatres

Power dissipation from motors, electrical equipment, and process work. The total thermal load arising from internal power dissipation depends upon the frequencies of use, power ratings, and efficiencies.

The efficiency of conversion of electricity to heat is very nearly 100%, most losses having taken place during the generation of this very high grade form of energy (cf. Chapter 1).

TABLE 6.10. CASUAL ADDED GAINS IN AN AVERAGE HOME

Type	Estimated frequency of use (h/year)	Rating (W)	Approximate equivalent continuous rating averaged over an annual period (W)
Thermal gain from hot-water tank heat leak (at 50 °C)	8760	50	50
Heat gain from use of hot water[a]	8760	50	50
Heat gain from cooking	1000	up to 3000	350
Heat gain from people (3)	4000	400	180
Heat gains from electrical appliances			
Lighting	700	800	65
Television	1000	200	25
Tape recorder	200	40	1
Record player	200	40	1
Radio	1000	40	5
Electric clock	—	—	Negligible
Automatic washing machine	300	1000	35
Spin drier	300	125	4
Tumble drier	300	1000	35
Iron	300	400	14
Refrigerator	8760	40	4
Freezer	8760	200	20
Cooker hood	1000	100	11
Electric kettle	100	3000	35
Toaster	50	1000	6
Food mixer	—	—	Negligible
Dish washer	500	750	42
Vacuum cleaner	300	250	9
Electric blanket	300	75	0.5
Electric shaver	—	—	Negligible
Hair drier	150	700	12
Battery charger	—	—	Negligible
Total			955

[a]It is assumed that all hot water used for baths, showers, hand washing clothes and dishes, etc., is rejected after dissipating 25% of its total heat to the interior of the building.

Table 6.10 lists estimated typical sundry heat gains for a domestic dwelling. It is seen that the annual energy release for non-heating purposes can amount to an equivalent of ~ 1 kW of continuous power dissipation.

6.4.4   **Annual heat surplus**

The final figures for the mean and total annual heat balance of the considered building are as follows:

	Total annual energy flow	
	Equivalent continuous power rating (kW)	Total energy (10^6 MJ per year)
Net gain from the external environment (neglecting cloud cover or other climatic modification)	3.9	0.12
Casual internal added gains	0.9	0.03
Ventilation requirements	−0.6	−0.019
Totals	4.2	0.131

Thus the building system under consideration enjoys an annual net heat surplus equivalent to a continuous power rating of 4.2 kW, and not a net heat loss as commonly believed. The need for 7.24 kW to offset transmission losses (cf. Table 5.6) could thus be avoided in sensible energy-conscious designs which would include solar energy collection, thermal storage, heat reclamation, and the proper application of techniques of thermal isolation together with controlled ventilation. In order to assess the design criteria at actual sites, local statistical meteorological data (including incidence and duration of clouds, mists, and fog) must be consulted to modify the theoretical figures obtained for total solar incidence.

Chapter 7

Thermal Network Analysis

7.1 COMPLEX THERMAL SYSTEMS

Previous chapters have indicated that the thermal analysis of even quite simple mechanical structures can be difficult and tedious. Complete solutions must specify temperature distributions, local heat fluxes, and overall heat flows for both steady and transient conditions. Complex thermal systems may, however, often be investigated with sufficient accuracy by replacing the thermal continuum by an array containing n discrete nodal points connected by thermal resistances, across each of which one-dimensional heat flow occurs. Thus linear temperature gradients are assumed to occur between each node. A thermal capacity, representative of the portion of the physical structure associated with each nodal point, is assigned to each node. Strictly, temperature distributions across system components are not linear. Thus the more nodes allocated to the system, the closer the internodal temperature gradients approach linearity and the more accurate the resulting solution.

A steady-state or transient heat balance equation is obtained for each node and a matrix of $(n - 1)$ equations is formulated. These may be solved if only two of the variables are known, using analytical, graphical, or approximate numerical techniques.

Thermal resistances and conductances $(= 1/R)$ which occur for different modes of heat transfer are summarised as follows:

For conduction: $Q = \dfrac{kA}{L} \Delta T,$ (7.1)

and so $\qquad R = \dfrac{L}{kA} \quad \text{and} \quad C = \dfrac{kA}{L}.$ (7.2)

For convection: $Q = hA\Delta T$ (7.3)

and so $\qquad R = 1/hA \quad \text{and} \quad C = hA.$ (7.4)

For radiation: $Q = \mathscr{F}\sigma A(T_1^4 - T_2^4)$ (7.5)

$\qquad\qquad \simeq h_r A(T_1 - T_2),$ (7.6)

where h_r is an equivalent heat transfer coefficient.

Thermal capacitance M is calculated from

$$M = mc_p = \rho V c_p.$$ (7.7)

7.2 STEADY-STATE NODAL ANALYSIS

7.2.1 Nodal equations

In practice, convective and radiative heat transfer coefficients are dependent upon system temperatures. In order to demonstrate the network approach to thermal analysis, however, it is assumed in the first instance that each nodal interconnecting resistance is constant and independent of temperatures and heat fluxes. More sophisticated computer techniques, based on iterative methods of solution, can easily handle temperature-dependent heat transfer coefficients.

Single node (Fig. 7.1a):
 Rate of heat entering node = rate of heat leaving node.   (7.8)

$$\dot{q}_1 = \dot{q}_1.$$

Series connection of two nodes (Fig. 7.1b):

For node 1: $\dot{q} = \dot{q}_1 = \dfrac{T_1 - T_2}{R_{12}}$

For node 2: $\dfrac{T_1 - T_2}{R_{12}} = \dot{q}_2 = \dot{q}.$

Thus one of the two equations describing heat transfer through the system is redundant. Only one equation is of use and there are three unknown quantities—T_1, T_2, and $\dot{q}$. Two of these unknowns must be specified in order to evaluate the third.

Series connection of n *nodes.* By performing a steady-state heat balance at each node in the system, n equations are deduced of which one is redundant. In general, there arise $(n-1)$ useful equations and $(n+1)$ unknown values. Thus if two values (i.e. two local temperatures, two local heat fluxes, or a local heat flux and any nodal temperature) are specified, the system can be solved for all local temperatures and heat fluxes.

Nodes in series and in parallel. Taking heat balances at each nodal point shown in Fig. 7.1c:

For node 1: $\dot{q} = \dfrac{T_1 - T_2}{R_{12}} + \dfrac{T_1 - T_3}{R_{13}}.$

For node 2: $\dfrac{T_1 - T_2}{R_{12}} = \dfrac{T_2 - T_4}{R_{24}}.$

For node 3: $\dfrac{T_1 - T_3}{R_{13}} = \dfrac{T_3 - T_4}{R_{34}}.$

For node 4: $\dfrac{T_2 - T_4}{R_{24}} + \dfrac{T_3 - T_4}{R_{34}} = \dot{q}.$

Four equations are produced, one of which is again redundant. Thus, as for the cases considered previously, there are $(n-1)$ equations and $(n+1)$ unknown values.

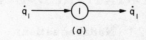

(a)

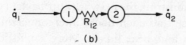

(b)

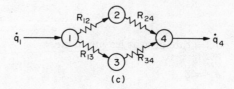

(c)

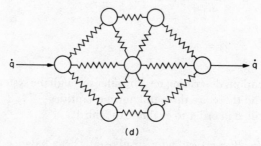

(d)

Fig. 7.1. Thermal networks: (a) for a single node; (b) for a series connection of two nodes; (c) for nodes in series and parallel; (d) a complex system.

Complex networks (**Fig. 7.1d**)

The general procedure for establishing the nodal equations is straightforward:

(i) A heat balance is performed at each node.

(ii) The resulting $(n - 1)$ equations contain $(n + 1)$ unknowns.

(iii) Two values must be specified to solve $(n - 1)$ equations for $(n - 1)$ unknowns.

Solving the network equations. Steady-state thermal networks containing constant resistances connecting n nodes yield $(n - 1)$ simultaneous linear equations. These are handled most easily using the Gaussian elimination matrix method of solution (see Appendix I [48]) or a basic relaxation technique [18].

7.2.2 **Steady-state conduction—finite difference method**

If thermal conduction alone takes place, as in the interior of a solid, the solid continuum is replaced by n nodal points in a regular rectangular array. Figure 7.2 shows part of a two-dimensional system. The nine nodes shown are separated by $\Delta x(\Delta y)$ and each has a volume $\Delta x \Delta y$ per unit depth assigned to it. The finite difference method of formulating equations [18]

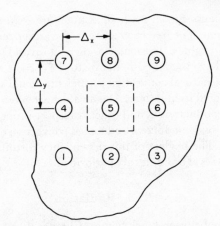

Fig. 7.2. Representation of part of a two-dimensional continuum by an array of nodal points.

connects adjacent nodes by thin solid bars each having a thermal resistance per unit depth of

$$\frac{L}{kA} = \frac{\Delta x}{k\Delta y} \quad \text{or} \quad \frac{\Delta y}{k\Delta x}. \tag{7.9}$$

a heat balance is then performed for each node resulting in a solvable set of linear equations provided that at least two of the temperatures are known. Techniques of solution include relaxation methods [18], Gauss–Siedal iteration [18], or Gaussian elimination (Appendix I). This method is the basis of the simple network approach to heat transfer through a solid wall as demonstrated in Chapter 5.

7.2.3   **Forced convection**

When rates of heat flow across solid boundaries are affected by forced convective mechanisms, a resistance between the boundary and the adjacent fluid may be added to the nodal network.

$$R = \frac{1}{hA}, \tag{7.10}$$

where the heat transfer coefficient h ($W\,m^{-2}\,K^{-1}$) depends upon the thermal conductivity of the fluid, the Reynold's number, and the Prandtl number (cf. Chapter 5), i.e.:

$$h = f(Re, Pr, x, k).$$

7.2.4   **Free convection**

Difficulties arise when the adjacent fluids are in natural convection because heat transfer coefficients then depend appreciably upon the temperature difference between the surface of the solid and that of the bulk fluid (cf. Chapter 4), i.e.

$$h = f(Gr, Pr, x, k)$$

and $$Gr = f(\Delta T, \text{transport properties}).$$

Thus the matrix which contained only constants for the conductive analysis contains functions of the unknown temperatures when free convection occurs. One possible method of solution, related to the Gauss–Siedel method, assumes approximate values for all temperature before commencement. The free convective heat transfer coefficients are calculated on the basis of these assumed values. The matrix is then set up and solved for the temperatures. The new values obtained are then used to iterate to the correct solution (cf. Chapter 5). In many instances, however, natural convection coefficients do not vary extensively in the temperature ranges encountered in buildings. Mean values are often used [19].

Strictly, the thermal conductivities and other transport properties of materials also vary with temperature. Thus solid conductive and forced convective heat transfer coefficients are temperature-dependent. The neglect of this dependency fortunately leads to insignificant error when thermal analyses of buildings are performed.

7.2.5 Radiation

For radiation interchange between black bodies (Chapter 2),

$$Q = \mathscr{F}_{12}\sigma A(T_1^4 - T_2^4), \tag{7.11}$$

where $\mathscr{F}_{12}$ is a configuration factor which varies from zero to unity depending upon the system shape and surface emissivities, and σ is the Stefan–Boltzmann constant $(=5.67 \times 10^{-8}$ W m^{-2} K$^{-4})$. If radiation occurs in the absence of any other mechanism of heat transfer the solution is relatively simple. The matrix

$$\{q\} = [K]\ \{T^4\}$$

is established and solved for $(T^4)_{1,n}$ and hence $T_{1,n}$ (see Appendix I). This technique can be extended to cope with grey-body thermal radiation interchanges without much difficulty[18].

7.2.6 Combined modes of heat transfer

When heat transfer by convection, conduction, and radiation occur together in a system, the coefficient matrix derived contains constants, terms in T, T^4, and T^n, where n is a fractional power. Because the equations are no longer linear, the elimination method of solution is no longer applicable. Two basic groups of solution remain: (a) iterative techniques, and (b) semi-iterative techniques using linear approximation.

Approximate method of solution for small temperature differences. Whilst the Gauss–Siedel iterative method is often recommended for general heat transfer problems which include radiative heat exchanges because the absolute temperatures encountered in buildings do not encompass a substantial range of values, the following approximation for radiant interchanges may be used to reduce the network system to linearity:

$$\begin{aligned}
\dot{Q} &= A\mathscr{F}\sigma(T_1^4 - T_2^4) \\
&= A\mathscr{F}\sigma(T_1^2 + T_2^2)(T_1 + T_2)(T_1 - T_2) \\
&\simeq h_r(T_1 - T_2).
\end{aligned} \tag{7.12}$$

Absolute temperatures generally range from $\sim 270°C$ to $\sim 300°C$.

Thus
$$h_r \simeq 93 \times 10^6\ \mathscr{F}\sigma = 5.8\ \mathscr{F} \tag{7.13}$$

and so
$$R \simeq (5.8\ \mathscr{F})^{-1}. \tag{7.14}$$

The configuration factor $\mathscr{F}$ may be calculated using eqn. (2.78).

Relative shape factors F_{12} for many simple and complex radiating systems are available in reference [49]. In building technology, the configurations most commonly occurring are parallel rectangles (for floor-to-ceiling interchanges, or for heat transfer between opposite walls) or adjacent rectangles (for abutting perpendicular walls) (Fig. 2.16). Emissitivities at room temperature are listed in Table 7.1.

TABLE 7.1. EMISSIVITIES OF SELECTED SURFACES AT ROOM TEMPERATURE [18]

Material	Emissivity	Material	Emissivity
Polished aluminium	0.04	White enamel	0.90
Weathered aluminium	0.40	White paper	0.95
Aluminium roofing	0.22	Plaster	0.91
Polished brass	0.10	Paints	0.95
Oxidised brass	0.61	Ice	0.97
Rough steel plate	0.94	Water	0.96
Weathered stainless steel	0.85	Wood	0.93
Asphalt	0.93	Glass	0.93
Red brick	0.93		

7.2.7 Demonstrations—steady-state system with constant resistances

EXAMPLE 1
The simple thermal network shown in Fig. 7.3 will be analysed.

Numbering nodes. To facilitate digital solution it is advantageous to obtain a well-banded matrix [50]. This may be achieved if, when numbering nodal points, the differences in numerical integer designations for adjacent nodes are arranged to be as small as possible.

Resistances and conductances. The resistances present in the network are numbered according to the nodes they connect, i.e. R_{ij}, connects nodes i and j. A reciprocity relationship applies at each resistance, i.e. $R_{ij} = R_{ji}$. Conductances between nodes are the reciprocals of the resistances, i.e. $C_{ij} = (R_{ij})^{-1} = C_{ji}$. Whilst establishing the system of governing equations it is preferable to use conductances rather than resistances. The notation ΔT_{ij} is adopted to represent $(T_i - T_j)$.

Nodal equations

Node	Heat balance equation
	(heat entering node = heat leaving node)
1	$\dot{q}_1 = C_{13}\Delta T_{13} + C_{14}\Delta T_{14} + C_{12}\Delta T_{12}$
2	$\dot{q}_2 = C_{31}\Delta T_{21} + C_{24}\Delta T_{24}$
3	$0 = C_{13}\Delta T_{13} + C_{34}\Delta T_{34} + C_{35}\Delta T_{35} + C_{36}\Delta T_{36}$
4	$0 = C_{14}\Delta T_{14} + C_{24}\Delta T_{24} + C_{34}\Delta T_{34} + C_{46}\Delta T_{46}$
5	$-\dot{q}_5 = C_{36}\Delta T_{36} + C_{46}\Delta T_{46} + C_{56}\Delta T_{56}$

or, in matrix form,

$$\begin{Bmatrix} \dot{q}_1 \\ \dot{q}_2 \\ 0 \\ 0 \\ -\dot{q}_5 \end{Bmatrix} = \begin{bmatrix} C_{12}+C_{13} & -C_{12} & -C_{13} & -C_{14} & & \\ +C_{14} & & & & & \\ -C_{21} & C_{21}+C_{24} & & -C_{24} & & \\ & & C_{31}+C_{34} & -C_{34} & -C_{35} & -C_{36} \\ & & +C_{35}+C_{36} & & & \\ -C_{41} & -C_{42} & -C_{43} & C_{14}+C_{24} & & -C_{46} \\ & & & +C_{34}+C_{46} & & \\ & & -C_{53} & & C_{35}+C_{56} & -C_{56} \\ & & -C_{63} & -C_{64} & -C_{65} & C_{36}+C_{46} \\ & & & & & +C_{56} \end{bmatrix} \begin{Bmatrix} T_1 \\ T_2 \\ T_3 \\ T_4 \\ T_5 \\ T_6 \end{Bmatrix}$$

Having constructed the matrix, it is worth while making the following observations [51].

Heat flux vector $\{\dot{q}\}$. The vector containing the heat flows $\{\dot{q}\}$ consists of component heat fluxes $\dot{q}_i$, where the numerical value of each term is the sum of the heat inputs to the node under consideration (inwards is positive). For all internal nodes, $\dot{q} = 0$.

Conductance matrix $[C]$. Diagonal values $C_{ii}(= C_{jj})$ are the sums of the values of conductance of all resistors connected to the ith node ("self-conductance").

Off-diagonal values $C_{ij}(i \neq j)$ are the negatives of the values of the conductances of the resistors connected between the ith and jth nodes ("mutual conductance").

The matrix is symmetrical about the leading diagonal because $C_{ij} = C_{ji}$. For a check procedure it is noted that the addition of numerical values in each column or row produces a residual value of zero.

Temperature vector. The component values of the temperature vector $\{T\}$ consist of the nodal temperatures T_i.

Simplified representation. The system of equations may be expressed simply in matrix notation as

$$\{\dot{q}\} = [C] \{T\}. \tag{7.15}$$

Nodal connectivities. The connectivity between nodes i and j is an integer which takes the value 1 when node i is connected to node j via an impedance, or zero when no connection exists. If connectivities and resistances are known, the matrix may be established without the necessity of performing each individual heat balance or even of constructing the thermal network diagram.

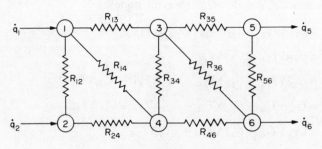

Fig. 7.3. Thermal network analysed.

Connectivity array. For the circuit under consideration, the array of connectivities is

	1	2	3	4	5	6
1	1	1	1	1		
2	1	1		1		
3	1		1	1	1	1
4		1	1	1		1
5			1		1	1
6			1	1	1	1

It is noted that the connectivity array is also symmetrical about the leading diagonal. Examples of connectivity arrays for well-banded and badly-banded numbered network systems are shown below.

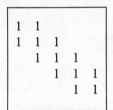

1	1			
1	1	1		
	1	1	1	
		1	1	1
			1	1

1					1
	1			1	1
		1		1	1
			1		
	1	1		1	
1	1	1			1

Well-banded system   badly-banded system

A conductance or resistance array may also serve as a connectivity array; where a value of conductance or resistance exists, a connectivity exists.

Governing matrix. Having determined the connectivities and the values of the conductances, the heat balance matrix is easily deduced. We shall, for the purposes of demonstration, assume arbitrary values of all conductances present in Fig. 7.3 of unity.

$$
\begin{Bmatrix} \dot{q}_1 \\ \dot{q}_2 \\ \dot{q}_3 \\ \dot{q}_4 \\ \dot{q}_5 \\ \dot{q}_6 \end{Bmatrix}
=
\begin{bmatrix}
3 & -1 & -1 & -1 & & \\
-1 & 2 & & -1 & & \\
-1 & & 4 & -1 & -1 & -1 \\
-1 & -1 & -1 & 4 & & -1 \\
 & & -1 & & 2 & -1 \\
 & & -1 & -1 & -1 & 3
\end{bmatrix}
\begin{Bmatrix} T_1 \\ T_2 \\ T_3 \\ T_4 \\ T_5 \\ T_6 \end{Bmatrix}
$$

To check the matrix: observe symmetry and check that the addition of numerical values in each row and in each column leaves zero residuals.

EXAMPLE 2

Figure 7.4 illustrates a two-dimensional representation of a simple building subjected to solar loading at the roof only. Heat losses occur by transmission through the walls and the base and internal radiation interchanges also occur. For simplicity, each boundary is assigned only one internal node. For more accurate results, at least four internal nodes are recommended when considering heat flows through the walls of buildings. It is assumed that the area of each boundary face is so large compared with its thickness that one-dimensional heat flows can be assumed.

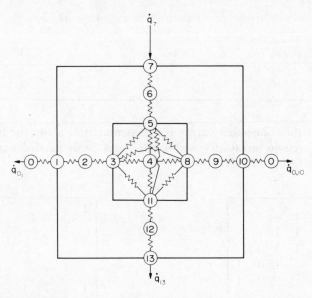

Fig. 7.4. The equivalent thermal network superimposed upon a sketch representing a two-dimensional section through a building.

Assigned values: T_o = 11°C (outside air temperature).

$\quad\quad\quad\quad\quad\quad T_4$ = 25°C (inside environmental temperature).

$\quad\quad\quad\quad\quad\quad T_{13}$ = 21°C $(T_o + 10°C)$.

$\quad\quad\quad\quad\quad\quad \dot{q}_7$ = 300 W m^{-2} (solar incident radiation).

Representative values for the conductive and convective plus radiative resistances are as follows:

$R_{0,1}$ = 0.4 $\quad\quad$ $R_{0,10}$ = 0.4

$R_{1,2}$ = 0.3

$R_{2,3}$ = 0.3

$R_{3,4}$ = 0.4 $\quad\quad$ $R_{3,5}$ = 0.9 $\quad$ $R_{3,8}$ = 0.9 $\quad$ $R_{3,11}$ = 0.9

$R_{4,5}$ = 0.4 $\quad\quad$ $R_{4,8}$ = 0.4 $\quad$ $R_{4,11}$ = 0.4

$R_{5,6}$ = 0.3

$R_{6,7}$ = 0.3

$R_{8,9}$ = 0.3 $\quad\quad$ $R_{8,11}$ = 0.9

$R_{9,10} = 0.3$

$R_{10,11} = 0.3$

$R_{12,13} = 0.3$

(all values being given in $m^2 \ K \ W^{-1}$).

These values have been calculated using the convective resistances estimated in Chapter 5 together with radiative resistances calculated using eqns. (2.78) and (7.6) and Fig. 2.16. The shape factor between the parallel walls is ~ 0.2 and that for the perpendicular walls also ~ 0.2. A surface emissivity of 0.93 was adopted and all surface areas were assumed to be equal. Then $\mathscr{F} = 0.194$ for all configurations present. The heat balance matrix in terms of conductances is presented below:

	0	1	2	3	4	5	6	7	8	9	10	11	12	13	
$\dot{q}_{0,1} + \dot{q}_{0,10}$	5	−2.5									−2.5				11
0	−2.5	5.8	−3.3												T_1
0		−3.3	6.7	−3.3											T_2
0		−0.3	−3.3	9.1	−2.5	−1.1			−1.1			−1.1			T_3
0				−2.5	10	−2.5			−2.5			−2.5			25
0				−1.1	−2.5	9.1	−3.3		−1.1			−1.1			T_5
0						−3.3	6.6	−3.3							T_6
300							−3.3	3.3							T_7
0				−1.1	−2.5	−1.1			9.1	−2.2		−1.1			T_8
0									−3.3	6.6	−3.3				T_9
0	−2.5									−3.3	5.8				T_{10}
0				−1.1	−2.5	−1.1			−1.1			9.1	−3.3		T_{11}
0												−3.3	6.6	−3.3	T_{12}
$\dot{q}_{13}$													−3.3	3.3	21

and $300 = \dot{q}_{0,1} + \dot{q}_{0,10} + \dot{q}_{13}$

There are 14 independent equations and 14 unknowns. Thus the system may be solved using relaxation, elimination, or iterative techniques. Computer subroutines in Fortran for the computation of matrix elements and for the solution of the resulting matrix are provided by Huelsmann [51].

7.3 UNSTEADY HEAT FLOW SYSTEMS

7.3.1 Introduction

Although the maximum intensity of solar radiation occurs at midday, the outside air temperature is highest during mid-afternoon. The time lag between these peaks occurs because the environmental air is warmed only by the surfaces with which it comes into contact—these surfaces being warmed directly by solar radiation. The thermal capacities of the surface substrates and the mass of the air delay instant temperature response. Thus building structures further dampen variations in incident heat fluxes, and so peaks in interior temperatures are also encountered some time after midday. It is not uncommon for the inside environmental air temperature to remain quite high for several hours after sunset. Thus, when heat is transferred to a system in thermal equilibrium, it absorbs thermal energy and its temperature increases. Some time elapses before transient effects disappear and steady-state conditions are regained. During this transition period—known as the transient state—the thermal content of the system changes according to its thermal capacity and the rates of heat flow across the boundary.

7.3.2 **Systems with negligible internal resistance**

Many transient heat flow problems can be solved with reasonable accuracy by assuming that the internal resistance of the subsystem is so small that the temperature within it is essentially uniform at any instant [18]. The Biot number (cf. section 2.5.4) relates internal and external resistances. The error introduced by assuming temperature uniformity is less than 5% when $Bi < 0.1$. Equation (2.93) describes the temperature–time response of a heating or cooling solid for systems with negligible internal resistance,

$$\frac{T - T_\infty}{T_o - T_\infty} = \exp\left[-\left(\frac{hA}{c\rho V}\right)t\right],$$

(2.93)

where $c\rho V/hA$ is the time constant Θ of the system, i.e. the time taken for the temperature difference between the system and its surroundings to change to 36.9% of its initial value (at time $t = 0$).

When V/A is replaced by L and the numerator and denominator of the time constant are each multiplied by Lk_s, eqn. (2.93) becomes

$$\frac{T - T_\infty}{T_o - T_\infty} = \exp(-Bi\,Fo),$$

(7.16)

where Bi, the Biot modulus $= hL/k_t$, and Fo is the Fourier modulus $\alpha t/L^3$ which contains the time-dependence of the system.

Composite systems or bodies containing n components may be analysed using a "lumped capacity" method of analysis [18] in which the overall time constant is calculated from

$$\left(\frac{c\rho V}{UA}\right)_{tot}^{-1} = \sum_{i=1}^{n}\left(\frac{c\rho V}{hA}\right)_i^{-1},$$

individual component time constants being defined by the individual terms $\left(\dfrac{c\rho V}{hA}\right)_i$.

7.3.3 **Periodic heat flow in systems with negligible internal resistance**

In many transient problems the temperature of the medium surrounding the system varies with time. Equation (2.92) then becomes

$$c\rho V dT = \hbar A_s(T - T_\infty)dt$$

in which T_∞ is dependent upon time, i.e. $T_\infty = T_\infty(t)$. The resulting linear non-homogeneous differential equation with constant coefficients, viz.

$$\frac{dT(t)}{dt} + \frac{\hbar A_s}{\rho c V}T(t) = \frac{\hbar A_s}{\rho c V}T_\infty(t)$$

(7.16a)

may not be solved by simply separating the variables (cf. section 2.5.4). The solution consists of the sum of two parts [18]:

(a) The *particular integral* which satisfies the complete equation and contains no arbitrary constants. Physically, this portion of the solution represents the temperature–time history of the system after transient phenomena have disappeared or the steady-state solution. $T(t)$ has the same functional form as $T_\infty(t)$.

(b) The *complementary function* which contains constants of integration whose values are obtained from the initial $(t = 0)$ or boundary conditions. The complementary function represents the transient response of the system temperature which arises because of a lack of initial equilibrium, and will decay exponentially as when T_∞ is constant.

The complete solution is the sum of these two parts. The boundary and initial conditions must always be applied to the complete solution and never to the transient portion alone.

7.3.4 Transient heat flow in an infinite plate

When the Biot modulus appertaining to a system is greater than 0.1, the simple lumped capacity approach is inapplicable and the solution requires that the general heat conduction equation is satisfied by a derived temperature distribution.

For one-dimensional heat flow in the absence of sources and sinks,

$$\frac{1}{\alpha}\frac{\partial T}{\partial t} = \frac{\partial^2 T}{\partial x^2} \quad \text{for } 0 \leq x \leq L. \tag{7.17}$$

Kreith [18] has demonstrated analytical techniques applicable for simple geometries (solid walls, long cylinders, and spheres) subjected to sudden changes of environmental temperature and has presented solutions for time-dependent temperature distributions and instantaneous heat fluxes in the form of dimensionless charts. Alternative graphical (i.e. the Schmidt plot) methods have greater range of application but may be applied to one-dimensional heat flow systems only. Numerical methods are generally recommended for greater accuracy in complex situations, especially when a high-speed digital computer is available.

7.3.5 Application to buildings

Parameters of interest to designers investigating the unsteady-state performance of buildings include *overall and component time constants*, and the *time lags* and *decrement factors* of components subjected to periodic heat flux variations. When a solid wall is subjected to a periodic heat input on one side (Fig. 7.5) the time lag Φ is the displacement between heat flux–time variations either side of the wall, whilst the decrement factor ∂ represents the reduction in amplitude from that of the input to that of the induced characteristic at the cooler boundary of the wall.

The procedure generally adopted whilst assessing peak transient thermal loads on buildings [52] is due to McKay and Wright [53]. This analytical solution uses the sol-air temperature to trigger the response. Characteristics relating to homogeneous solid walls with one-dimensional heat flows have been derived [21, 52, 53]. More complex situations must be analysed using approximate numerical network techniques.

7.3.6 Numerical network analysis for unsteady-state systems

The internal temperature distribution in a steady-state system can be obtained by solving a matrix of coefficients describing a set of simultaneous equations. In transient systems, the initial temperature distribution is known, but its variation with time must be determined. It is therefore necessary to deduce the temperature distribution at some future time from a given set of initial data. When the approximate network technique is used, the system is divided into an array of nodes as for steady-state analyses. For the one-dimensional unsteady system

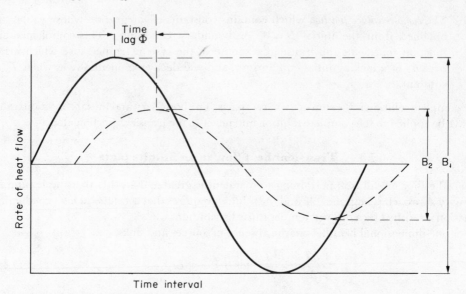

Fig. 7.5. Definitions of time lag and decrement factor for the periodic transfer of heat through a solid wall: ———— heat input; — — — heat output. Decrement factor $\partial = B_2/B_1$.

shown in Fig. 7.6), node 2, with a thermal capacity mc_p, is receiving heat from a source at a temperature T_1 via a resistance R_{12} and losing heat to node 3 through a resistance R_{23}. Because some heat is being stored in the capacitance associated with node 2, the instantaneous rate of heat flow from node 1 to node 2 does not equal the instantaneous rate of heat flow from node 2 to node 3 during transient adjustment. The governing heat balance equation is as follows: Heat entering node 2 in time dt = heat leaving node 2 in time dt + heat stored in node 2 in time dt, or, using the notation developed earlier and rearranging,

$$0 = \frac{\Delta T_{12}}{R_{12}} + \frac{\Delta T_{23}}{R_{23}} + (mc_p)_2 \frac{dT_2}{dt}, \tag{7.18}$$

or, in terms of conductances,

$$\{0\} = (C_{21} + C_{23})T_2 - C_{12}T_1 - C_{32}T_3 + M_2 T_2'', \tag{7.19}$$

i.e.

$$\{0\} = [-C_{12}(C_{21} + C_{23}) + M_2 - C_{32}] \begin{Bmatrix} T_1 \\ T_2 \\ T_2' \\ T_3 \end{Bmatrix} \tag{7.20}$$

where $T_2' = dT_2/dt$ and $M_2 = (mc_p)_2$.

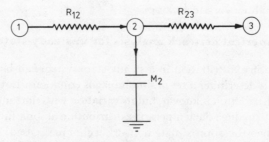

Fig. 7.6. Basic module of a one-dimensional unsteady heat flow system.

Thus $n - 1$ first-order differential equations will result for n nodes. These may be solved using techniques similar to those employed for solving linear equations if a substitution of the type $T = ke^{st}$ is made, i.e. $dT/dt = kse^{st}$.

The matrix of coefficients will then contain the variable (ke^{st}) which, being common to all terms, may be included in the temperature vector. The determinant of the matrix of coefficients is the characteristic equation of the system [57]. When this is equated to zero, the values of s which satisfy this condition are the natural frequencies of the system, and hence the reciprocals of the time constants. A solution to the matrix equation arises for each set of values of s. The addition of individual solutions provides the full solution for the instantaneous temperatures and heat fluxes.

Linearised solution. Alternatively, the first-order differential dT/dt may be replaced by the ratio of the approximate elements ΔT a:d Δt. Then eqn. (7.18) becomes

$$0 = C_{12}\Delta T_{12} + C_{23}\Delta T_{23} + M_2 \frac{\Delta T_2}{\Delta t}. \tag{7.21}$$

This is now a linear equation containing the time-dependent temperature step ΔT_2. Rearranging eqn. (7.21),

$$\Delta T_2 = -\left(\frac{C_{12}\Delta T_{12}\Delta t + C_{23}\Delta T_{23}\Delta t}{M_2}\right) \tag{7.22}$$

or

$$T_{2,t+1} = T_{2,t} - \frac{C_{12}\Delta T_{12,t} + C_{23}\Delta T_{23,t}}{M_2/\Delta t}. \tag{7.23}$$

Thus if the time interval and initial steady-state situation is known, the temperature at a later time can be estimated from a linear approximation. In a complex system each node produces a transient equation. It is to be noted that these equations do not have to be solved simultaneously, an equation of the type (7.23) representing a forward prediction.

Finite differences and Schmidt plots. The proceeding analysis forms the basis of finite difference solution techniques for three-dimensional heat flows in solids [18]. In general, eqn. (7.23) may be expressed as

$$T_{n,n-1} = \frac{1}{B}\left(T_{n+1} + T_{n-1}\right)_t + \left(1 - \frac{2}{B}\right)T_{n,t} \tag{7.24}$$

where $B = \Delta x^2/\alpha\Delta t$ and $\Delta x(= \Delta y = \Delta z)$ is the distance between adjacent nodes. The Schmidt plot [18] is a graphical method for solving two-dimensional systems (B in eqn. (7.24) is usually assigned the value 2 in order to reduce the equation to a more easily handled form). It must be emphasised that both these techniques are founded upon linear approximations. Thus the temperature gradients connecting nodes are assumed to be linear. The greater the number of nodes the more accurate the solution. Intricate thermal systems may be handled most easily using digitised data or by constructing an analogous electrical network [54, 55].

Generation of governing equations
  In general,

$$T_{t,t+1} = T_{i,t} - \frac{\sum\limits_{j=1}^{n}(C\Delta T)_{t,ij(j \neq i)}}{M_i/\Delta t} \tag{7.25}$$

or

$$T_{t,t+1} = T_{i,t} - \frac{\sum\limits_{j=1}^{n}(CT_i)_{t,ij(j \neq i)} - (CT_j)_{t,t=1,n(j \neq i)}}{M_i/\Delta t}. \qquad (7.26)$$

EXAMPLE 3

It is assumed that node 2 (described in example 2) has associated with it a thermal capacity per unit heat flow area of $M_2 = 10^5\,\mathrm{J\,m^{-2}\,K^{-1}}, i = 2, j = 1, 3,$

$$T_{2,t+1} = T_{2,t} - \left[\frac{\sum\limits_{3}^{1}(CT_2)_{t,2j(j=1,j=3)} - (CT_j)_{t,(j=1,j=3)}}{M_2/\Delta t}\right]$$

or

$$T_{2,t+1} = T_{2,t} - \frac{\Delta t}{M_2}\left[(C_{21} + C_{23})\,T_2 - C_{21}T_1 - C_{23}T_3\right]. \qquad (7.27)$$

From example 2, $C_{12} = C_{23} = 3.3\,\mathrm{W\,m^{-2}\,K^{-1}}$, and so

$$T_{2,t+1} = T_{2,t} - \frac{\Delta t}{10^5}\,(6.6T_2 - 3.3T_1 - 3.3T_3).$$

The value of the interval Δt must be selected so that it is less than the smallest component time constant in the network. In the present case, the only time constant in the system exists at node 2. For this node,

$$\text{time constant} = M_2/\textstyle\sum C_{2j}$$
$$= 10^5/6.6$$
$$= 1.5 \times 10^4\,\mathrm{s} = 4.2\,\mathrm{h}.$$

Thus a time constant of 1.5×10^3 s can be selected to facilitate calculations.

Initial temperature estimates are assumed as follows:

$T_{1,0} = 50°\mathrm{C}$ (assumed remaining constant),

$T_{2,0} = 25°\mathrm{C},$

$T_{3,0} = 25°\mathrm{C}$ (assumed remaining constant).

Then eqn. (7.27) becomes

$$T_{2,t+1} = 0.9\,T_{2,t} + 3.75$$

(i.e. a steady-state condition is achieved when $T_{2,t+1} = T_{2,t} = 37.5°\mathrm{C}$).

Response

Time (h)	T_2 (°C)
0	25
0.42	26.25
0.84	27.37
1.26	28.40
1.68	29.31
⋮	⋮
	37.50

In a real system, all nodal temperatures will change after each time interval.

Notes on the application of transient network techniques. Example 3 demonstrates how the transient response to abrupt boundary temperature changes may be deduced. If the external temperatures or boundary heat fluxes vary periodically, instantaneous values may be assigned to the nodes representing the external temperatures or the nodes on the boundary may be given instantaneous heat flux inputs. Thus decrement factors, time lags, temperature–time responses, and instantaneous heat flow through each node may be easily deduced for multinodal systems.

Chapter 8

Energy Thrift

8.1 FUEL PRICES AND CONVERSION EFFICIENCIES

After estimating the annual energy requirements of a building, a heating or cooling system must be selected. Taking 1976 prices as a guide and using electricity prices as a datum, the relative percentage costs of energy delivered to the consumer are as follows:

Electricity (on-peak)	100%
Coal	16%
Fuel oil	14%
Gas	14%

The cost per intrinsic MJ of energy delivered in the basic fuels is approximately one-sixth of that available in electricity. The latter, however, is almost 100% efficient in conversion to thermal energy, nearly all conversion losses having occurred at the generating station, where ~70% of primary energy is dissipated as waste heat during the transformation from fossil fuel to mechanical energy. Thus the 100% of energy equivalent indicated for electricity really represents an energy content in the primary fuel of 303%!

The efficiencies of conversion from chemical energy in fossil fuels to thermal energy range from 25% to 85% depending upon the design, installation, and maintenance of the conversion device. Thus, in terms of useful thermal energy provided, the relative costs in monetary and useful energy terms of systems fuelled by various energy forms are as follows:

	£/MJ of energy	Relative cost of useful energy provided
Electricity	100%	100%
Coal	18–64%	12–45%
Fuel oil	16–56%	11–39%
Gas	11–39%	11–39%

It is seen that, even after taking conversion efficiencies into account, the running costs of coal, oil, and gas-fired systems are at least half those of electrical heating installations unless off-peak electricity is available at a discount. There are, of course, other factors which affect the choice of fuel. Relative prices vary across the country, and, indeed, some fuels may not be available everywhere. Electrical systems are cleaner, simple to install and maintain, and respond almost instantaneously to regulation. Off-peak systems have the additional advantage of heating buildings during the night at little extra cost. Oil and solid fuels require storage space and transportation. Fossil-fuelled devices are relatively complex and need regular servicing. Because of long response delays, solid fuel systems, and, to a lesser extent, gas and oil-fuelled arrangements, experience wastage of heat during periods when output is not required. From the energy conservation viewpoint, space heating should not be accomplished using electricity, but rather via the direct use of fossil fuels, by the utilisation of solar energy, or by waste heat recovery from power stations.

8.2 REDUCTION OF ENERGY CONSUMPTION

All energy-consuming appliances and systems should be designed and selected on the basis of energy running costs, coupled with long-life criteria, rather than upon the basis of initial capital cost. Historically, fossil fuels have been relatively cheap, and so open fires in conjunction with higher levels of clothing have been used to achieve human thermal comfort in poorly insulated buildings. As energy sources become scarcer, and so more expensive, energy flows must be investigated and critically examined so that wastage areas may be identified and savings may be achieved.

Energy-saving measures can be grouped into three categories:

(i)   *Methods applicable during architectural design*
  (a) Optimising building size and shape.
  (b) Optimising the location and orientation with respect to the predominant solar influx and prevailing wind direction.
  (c) Optimising the ratio of glazed to solid boundaries.

(ii) *Simple low-cost methods applicable to existing buildings* (in order of cost effectiveness)
  (a) Reduction of room temperatures.
  (b) Draughtproofing.
  (c) Application of thermal insulants.
  (d) Installation of double glazing.

(iii) *More complex and costly measures* (in approximate order of cost effectiveness)
  (a) Matching thermal response with controls.
  (b) Reducing influx to minimum ventilation requirements.
  (c) Introducing more efficient energy distribution systems.
  (d) Utilising solar or other alternative energy.
  (e) Using heat recovery techniques.
  (f) Introducing a heat pump.

8.3 OPTIMISING BUILDING DESIGN FOR ENERGY SELF-SUFFICIENCY

An optimisation with respect to the shape and glazing ratios for multistorey rectangular-plan buildings oriented NS–EW and subjected to annual mean environmental conditions including solar influx will be attempted.

A detached flat-roofed building of height b with plane rectangular surfaces (Fig. 8.1) has walls of lengths a facing north and south, and walls of length c facing east and west. The north wall, the south wall, and the roof contain fractional areas of glazing, x'_N, x'_S, and x'_{rf} respectively, whilst the east and west walls each contain a fraction x'.

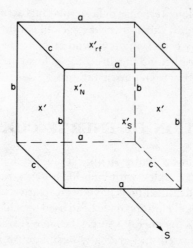

Fig. 8.1. Sketch of building shape to be optimised with respect to net heat gains.

At any instant of time, the net steady-state heat gain Q to the structure is given by

Q = solar gains through north, south, east, and west glazing plus solar gain through the roof glazing

   —transmission losses through the glazing

   —transmission losses through the walls

   —transmission losses through the roof

   —transmission losses through the base.

(The time lags for solar gains through the solid opaque components are neglected.)

If the overall heat transfer coefficients are U_{gz}, U_w, U_{rf}, and U_{bs} for the glazing, walls, roof, and base respectively,

$$Q = \sum SAI - \sum UA\Delta T, \tag{8.1}$$

where S represents the solar gain factor applicable for the glazing and I (W m^{-2}) the intensity of solar radiation in directions normal to the surfaces.

For transmission loss through the glazing: $\Delta T = T_i - T_o = \Delta T_o$, where T_i and T_o are the inside and outside temperatures.

For transmission loss through the walls: $\Delta T = T_i - T_{sa} = \Delta T_{sa}$, where T_{sa} is the outside environmental sol-air temperature.

For transmission loss through the base: $\Delta T = T_i - T_o + 10°C$.

Thus

$$Q = S\,(abx'_N I_N + abx'_S I_S + bcx'_E + bcx'_W + acx'_{rf}I_{rf})$$
$$- U_w[ab(1 - x'_N)\Delta T_{N,sa} + ab(1 - x'_S)\Delta T_{S,sa} + bc(1 - x')\Delta T_{E,sa} + bc(1 - x')\Delta T_{W,sa}]$$
$$- U_{rf}[ac(1 - x'_{rf})\Delta T_{rf,sa}]$$
$$- U_{gz}(abx'_N\Delta T_o + abx'_S\Delta T_o + bcx'\Delta T_o + bcx'\Delta T_o)$$
$$- U_{gz}(acx'_{rf}\Delta T_o)$$
$$- U_{bs}[ac(\Delta T_o + 10)]. \tag{8.2}$$

Over an annual period the mean heat gain to a building maintained at 25°C is (using values for I, ΔT_{sa}, and ΔT_o from Chapter 6)

$$\begin{aligned}
\dot{Q} = \; & S[(1238x'_N + 5199x'_S)ab + 6872x'bc + 5492x'_{r\,f}ac] \\
& - U_w(21.1-12.4x'_N-8.7x'_S)ab + 19.4(1 - x')bc \\
& - U_{rf}[12.4(1 - x'_{rf})ac] \\
& - U_{gz}[14ab('_Nx + x'_S) + 28x'bc] \\
& - U_{gz}(14x'_{r\,f}ac) \\
& - U_{bs}(4ac)
\end{aligned} \qquad (8.3)$$

A building which requires no net annual internal heating or cooling must be designed so that $\dot{Q} = 0$. To achieve internal thermal comfort it is also desirable that all internal surfaces are at the same temperature. Thus temperature drops through each wall and through the roof should also be identical as far as is possible at each instant of time. These criteria can be achieved by the proper selections of x', U, a, b, and c by introducing a degree of control using variable resistance boundary insulations and by incorporating a thermal store. Because eleven independent variables are present in eqn. (8.3) there is thus no unique optimum design with respect to energy flows through a building structure when solar gains are considered. If, however, the fractional areas of glazing and the U-values are preset, then an optimum shape can be deduced. Alternatively, given the shape of a structure, the U-values and glazing ratios can be optimised. Expressions of the form of eqn. (8.3) also serve when making comparisons among different conventional designs but often require computer evaluation. Arrangements incorporating solar energy collectors and thermal storage may be assessed using the general heat balance equation. Instantaneous thermal situations, and hence the transient behaviour of thermal systems, should also be investigated when designing control systems.

8.4 LOW-COST ENERGY-SAVING METHODS

If sufficient capital outlay is expended on energy-saving equipment during the construction of a new building, the resulting system can be self-sufficient in energy for subsequent space heating requirements. Unfortunately, most energy conservation hardware, due mainly to a lack of demand for it, is capital-intensive, and applications to existing buildings often require structural alterations. Consequently, because of a shortage of investment capital, government-initiated exhortations to save energy tend to encourage predominantly low-cost, instant return methods.

8.4.1 Reduction of room temperatures

The selection of optimum internal air temperatures for thermal comfort has been discussed in Chapter 3. It was seen that, because of social requirements, an inside environmental temperature of 25°C appears to be desirable for thermal comfort. The sensation of human thermal comfort is affected not only by air temperature but also by radiant temperatures, air velocities, humidities, the level of clothing, and metabolic activity. All these factors combine to produce a thermally comfortable state. If the air temperature is reduced, the occupants must either tolerate a thermally uncomfortable condition or must compensate by adjusting other factors, usually the level of activity or the amount of clothing.

A comparison of recommended dry-bulb temperatures desirable for human comfort [24], resulting from studies of subjective responses, exhibits differences between values obtained in successive measurements. Acceptable temperature criteria for thermal comfort in the United States [56] rose steadily from $19.5 \pm 1.5°C$ in 1900 to $25 \pm 1°C$ by 1960. The new ASHRAE comfort chart (Fig. 3.1) recommends a range between $24.5°C$ and $26.5°C$. This trend indicates the increased comfort expectations from room-conditioning systems and the fashion of wearing lightweight clothing throughout the year.

To achieve thermal comfort, the outer surface of the skin must be maintained at about $33°C$. An average effective boundary conductance due to heat losses from a nude body by radiation to surrounding surfaces and natural convection to the environmental air can be estimated as $U = 10 \ W \ m^{-2} \ K^{-1}$ ($\sim 5 \ W \ m^{-2} \ K^{-1}$ for convection plus $\sim 5 \ W \ m^{-2} \ K^{-1}$ for radiation). The emissivity of clothing on a substrate has been found to be less [41] than that usually assigned to human skin (~ 1.0), and so the presence of clothing could reduce the radiative heat transfer coefficient and hence the overall coefficient to $\sim 7.5 \ W \ m^{-2} \ K^{-1}$.

Assuming that the mean radiant temperature of the environment is equal to the air temperature, the required difference between the skin surface temperature and the environment for a nude body to dissipate a near-basal metabolic rate of $50 \ W \ m^{-2}$ is given by

$$\Delta T = \dot{q} R_{c+r},$$ (8.4)

where R_{c+r} is the combined surface resistance due to convective and radiative heat transfer ($\equiv 1/U \simeq 0.1 \ m^2 \ K \ W^{-1}$.

Thus

$$\Delta T = 50 \times 0.1 = 5°C.$$

An environmental temperature of $\simeq 28°C$ would therefore be required for comfort if sensible heat exchanges alone were considered.

Even within the comfort zone, however, it has been noted [37] that $\sim 25\%$ of the metabolic heat is rejected by respiration and the diffusion of water through the outer layers of the skin. Assuming a constant environmental relative humidity, the temperature difference needed to reject the remaining 75% from a naked body is reduced to $3.75°C$, resulting in a higher required environmental temperature of $29.25°C$. A lower mean temperature would result in net heat losses from the body, whereas a higher temperature would not allow the body's heat to be rejected without the onset of vaso-regulatory processes and active sweating.

The environmental temperature required for thermal comfort of a clothed person depends upon the level of clothing insulation and metabolic activity. Typical metabolic rates vary from $40 \ W \ m^{-2}$ whilst sleeping to $400 \ W \ m^{-2}$ during heavy work. If the comfort condition is defined as a neutral state in which the body needs to take no particular involuntary (i.e. sweating or shivering) or voluntary (i.e. adjustment of environment) action to maintain its thermal balance [24], it is possible to develop a simple model to obtain an estimate of the effects of clothing and its thickness on room temperatures required for thermal comfort.

The temperature difference ΔT between the skin's surface and the environment required to dissipate a metabolic rate $\dot{q}$ from a clothed body is given by

$$T = \dot{q}(R_{c+r} + R_{cl}),$$ (8.5)

where R_{cl} is the resistance of the clothing layer.

It has been found experimentally [41] that the thermal resistance of most common clothing assemblies are, to a first approximation, proportional to their effective uncompressed thicknesses δ, i.e.

$$R_{cl} = 23\delta.$$ (8.6)

Using eqn. (8.6), $R_{c+r} = (U_{c+r})^{-1} = 0.133 \text{ m}^2 \text{ K W}^{-1}$, and $T_s = 33°\text{C}$. Thus

$$T_o = 33 - \dot{q}(23\delta + 0.133) \tag{8.7}$$

and, applying the correction for latent heat losses, gives

$$T_o = 33 - 0.75 \, \dot{q}(23\delta + 0.133). \tag{8.8}$$

This relationship is plotted in Fig. 8.2.

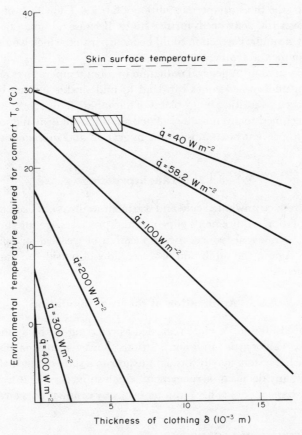

Fig. 8.2. Approximate effects of the level of clothing and the degree of activity upon acceptable room temperatures for human thermal comfort. The shaded region represents the comfort zone illustrated in the 1972 ASHRAE comfort chart.

The characteristics show the dramatic effect of clothing insulation on required room temperatures necessary to achieve thermal comfort. The ASHRAE comfort chart is intended to apply for a sedentary activity (with a corresponding $\dot{q} = 58.2 \text{ W m}^{-2}$) with a clothing insulating effectiveness of ~ 1 tog ($= 0.1 \text{ m}^2 \text{ K W}^{-1}$) corresponding to a thickness of ~ 4.5 mm. The region over which these limitations apply (Fig. 8.2) agrees with the derived characteristics. An increase in metabolic activity to 100 W m^{-2} (i.e. "light housework") would permit a reduction in acceptable room temperatures by about 9°C. Doubling the clothing insulation (i.e. to 10 mm thickness), whilst maintaining basal activity, allows room temperatures to be reduced by $\sim 5°\text{C}$. Such a reduction in temperature would save approximately

36% of annual heating costs when the mean outside environmental temperature is 11°C. Obviously, much greater savings can be accrued by the intelligent manipulation of clothing insulation and inside environmental temperature with respect to transient metabolic activity and frequency of occupation. The analysis neglects assymetrical heating or cooling from, for example, radiant sources or cold windows, and hence the radiant configuration factor between the body and surrounding surfaces. These factors should be considered in a more sophisticated analysis.

The correction for sweat diffusion and evaporation may be improved by introducing appropriate data for the moisture permeabilities of fabrics. The effects of clothing distribution and pressure over the body needs further study. Because a human body is usually only about 90% clothed, a simple correction could be incorporated which would slightly increase the required environmental temperatures.

Finally, a note of warning! A general reduction in room temperatures and the omission of space heating from unoccupied zones can lead to undesirable effects. Increased internal condensation will occur, resulting in problems of dampness. This aspect is further discussed in Chapter 9. Unoccupied rooms must be isolated against both heat and mass transfers from the heated spaces to prevent heat and vapour migration and subsequent condensation on cooler surfaces.

8.4.2 **Draughtproofing**

Draught excluders reduce unwanted cold and warm air infiltration. Care must be taken that the minimum ventilation requirements of people and processes are supplied. If a controlled ventilation system becomes necessary, the application of weather-stripping can prove an expensive exercise. Again, the inhibition of a free flow of outside air can cause increased condensation problems.

8.4.3 **Application of thermal insulation**

The term "thermal insulation" is often regarded as referring to the application of thermal insulants (i.e. cavity-wall or attic insulation). Strictly, however, thermal insulation embraces all methods by which a system may be brought nearer to a state of thermal isolation from its environment. Thus, any device or arrangement which suppresses heat transfer is a thermal insulator. The four basic modes of heat transfer occur by conduction, convection, radiation, and mass transfer.

8.4.3.1 CONDUCTIVE INSULATORS

Conduction of heat takes place in solids and in stagnant fluids and may be inhibited by reducing the number of solid bridges across a boundary by (a) reducing areas for heat flow, (b) increasing the lengths of the thermal paths, and (c) using low conductivity materials. Typical values for the thermal conductivities of solids, liquids, and gases are provided in Appendix IV. Stagnant gases have low thermal conductivities. Thus most thermal insulants trap air in interstices between solid particles. The insulating effectiveness is often proportional to the amount of fluid contained within an insulant. Liquids have higher thermal conductivities ($\sim$ an order of magnitude greater than gases and vapours), and so wet insulants result in much higher heat transfer rates. When two solid surfaces are pressed together, as a result of the very small real area of solid contact produced [57], the interfacial contact resistance is many times greater than the solid resistance of the abutting materials [58]. Stacks of thin, hard layers are often employed as superinsulators [59].

8.4.3.2 CONVECTIVE INSULATORS

The exposed areas from which convective heat transfer occurs can be reduced to inhibit convective losses (i.e. changes in human posture, such as folding the arms to reduce heat losses). For a given fluid, internal convection movements may be minimised by reducing fluid velocities or by evacuating the system. Natural convection is suppressed when $Gr \lesssim 2000$. The dimensions of heat flow channels can be reduced to obtain lower values of Gr. Thus the smaller a cavity separation, the less convection which ensues. Heights of cavities may also be reduced but not at the expense of introducing further solid transmission bridges. Draughts and stack effects should be avoided. Powders, foams, and fibres all serve as low-conductivity insulants by suppressing convection using small interstitial pores. Cellular pockets on surfaces, curtains, blinds, shutters, and double glazing also inhibit convective movements.

8.4.3.3 RADIATIVE INSULATORS

Even when convection is completely absent from a system, heat transfers still occur by radiation. This mode of heat transfer can take place in a vacuum requiring no intervening transport substance. Radiative energy transport may be minimised by reducing surface areas, by introducing highly polished, low emissivity shields and bends to prevent different temperature surfaces "seeing" each other, and by the use of low emissivity surfaces wherever possible. The rougher a surface, the nearer it resembles a blackbody radiator because radiation is absorbed during inter-asperity multiple reflections [60]. Roofs designed to exclude solar radiation are more efficient for this purpose when painted matt-white rather than when silvered because the surface equilibrium temperature depends upon the balance between emission and absorption.

8.4.3.4 MASS TRANSFER INSULATORS

The transport of heat within a displaced fluid is akin to convection but can also refer to thermal energy exchanges resulting from infiltration—these may be reduced by adhering to basic ventilating requirements. Rates of heat transmission associated with melting, evaporation, condensation, and solidification are often very much greater than rates of heat transfer resulting from any other mechanism [23]. High heat transfer rates across very small temperature gradients can be accomplished when changes of phase occur (such as condensation upon windows or walls, or in the evaporation of a liquid from a wet insulant). Any insulated system must be designed so that phase changes do not occur at the boundaries. Hence vapour barriers are sometimes used to contain condensable vapours within a system at temperatures above the prevailing dew-points (or to exclude vapour ingress from the external environment). The resulting humidity of the internal environmental air then becomes greater and often either increased ventilation or air conditioning becomes necessary.

8.4.4 Thermal insulants

A thermal insulant is designed [61] so that the thermal paths through the solid matter are long and circuitous to impede heat transmission by solid conduction. Fibres are often arranged in a direction perpendicular to the direction of the heat flow. The solid material should also be sufficiently opaque and reflective to reduce heat transfer by radiation, and the fluid pockets must be small enough to suppress convection. Properties of some common insulants are given in Appendix IV.

8.4.5   **Cavity walls**

The air gap introduced into a solid boundary by the inclusion of a cavity wall provides the major additional thermal resistance. The smaller the gap, the less inter-cavity convection that takes place. Radiant and conductive heat transmission through the fluid still occur. When the dimensions of the gap are such that the Grashof number $\simeq$2000 natural convection is completely suppressed and most heat passes across the cavity by gaseous conduction which increases if the gap is made smaller. There exists, therefore, an optimum inter-cavity separation ($\sim$ 19 mm for a vertical cavity) for minimum heat transference. An inter-cavity radiation shield will further increase the thermal resistance. The introduction of a cavity-fill insulant reduces thermal transmission by conduction, convection, and radiation. Vapour barriers should be incorporated to prevent vapour movements and condensation within the wall.

8.4.6   **Double-glazing**

Double-glazed windows cut down heat losses by impeding convective air movements within the cavity formed and by introducing an extra resistance to radiant transmission between internal and external environments. Noise is also suppressed, but it should be appreciated that, whereas the optimum pane separation for minimum inter-cavity heat transfer is of the order of 20 mm, the optimum for noise suppression is far greater than this. Infiltration is alleviated and, because the inside surfaces of the inner panes are warmer than the inner surfaces of single glazing, less condensation occurs on the glass. It must again be emphasised that the resulting increased moisture content of the room air must be offset by other means to prevent extra condensation occurring in, or on the surfaces of, the inside walls.

8.4.7   **Effects of applying low-cost energy-saving methods to buildings**

In Chapter 5, a building of dimensions 10 m by 10 m by 10 m, containing 30% glazing on all its façades, was evaluated in terms of steady-state heat losses over an annual period. This structure will now be further considered to ascertain the effects of introducing various methods of insulation.

8.4.7.1   INTER-CAVITY INSULANTS
The solid walls consist of two layers of 100 mm thick brickwork separated by an unventilated 50 mm wide air cavity. The overall thermal resistance of this structure has been calculated as 2.08 m^2 K W^{-1}. Table 8.1 lists values for overall resistances, overall heat transfer coefficients, and modified heat losses resulting when radiation shields or inter-cavity insulants are introduced.

It may be seen that the width of the cavity is such that very little heat transfer occurs by convection and so little is to be gained by attempting to suppress convection. Providing reradiation can be inhibited, inter-cavity shielding can reduce heat losses significantly. An inter-cavity insulant prevents convective movements and also acts as a radiation shield. When other resistances present in the thermal circuit are included, the overall resistance achieved is nearly independent of the type of insulant used.

8.4.7.2   DOUBLE-GLAZING
The installation of a second pane of glass introduces an air gap between the layers. If the width of this gap is optimised ($\sim$ 19 mm [62]) so that minimum intercavity heat transference

TABLE 8.1. EFFECTS OF INTER-CAVITY INSULATION

Mode of insulation	Thermal resistance of cavity (m^2 K W^{-1})	Overalll resistance[a] (m^2 K W^{-1})	Overall heat transfer coefficient (W m^{-2} K^{-1})	Heat loss[b] (kW)
(a) 50 mm air-filled cavity with two layers of 100 mm brickwork	1.02	2.08	0.48	1.88
(b) As (a) but with a thin radiation shield (neglecting reradiation)	2.04	3.10	0.33	1.29
(c) As (a) but with convection in the air gap completely suppressed	1.01	2.07	0.48	1.89
(d) As (b) but with convection in the air gap completely suppressed	2.00	3.06	0.33	1.25
(e) As (a) but with a glass fibre fill	1.35	2.48	0.41	1.63
(f) As (a) but with an expanded polystyrene fill	1.40	2.46	0.41	1.59
(g) As (a) but with a mineral wool fill	1.50	2.56	0.39	1.53

[a] An internal surface resistance of 0.4 m^2 K W^{-1}, a resistance for the solid brickwork of 0.66 m^2 K W^{-1}, and an external resistance of zero have been used to calculate the overall resistance. Radiation exchanges inside the cavity have been assumed to be of the order of the convective exchanges.

[b] A wall area of 280 m^2 has been adopted.

ensues, the additional series resistance associated with this air gap is of the order of 0.8 m^2 K W^{-1}. The total resistance (0.49 m^2 K W^{-1} for the single pane) then becomes 1.3 m^2 K W^{-1}, the overall thermal conductance U 0.77 W m^{-2} K^{-1}, and the heat loss through the glazing (120 m^2) is reduced from 3.36 kW to 1.3 kW. This substantial reduction demonstrates the increased cost effectiveness of double-glazing installations when large areas of glazing are architectural features.

8.4.7.3   CURTAINS
A heavy pleated curtain with an effective thickness of 10 mm has a thermal resistance of $\sim$0.23 m^2 K W^{-1} [41]. A 250 mm air gap provided between the curtain and the glazing has a further thermal resistance of $\sim$0.5 m^2 K W^{-1}. Thus the overall thermal resistance of a window consisting of a single pane of glass but covered by such a curtain is 1.22 m^2 K W^{-1}, the overall heat transfer coefficient U is 0.82 W m^{-2} K^{-1}, and the steady-state heat loss is reduced from 3.36 kW to 1.38 kW. Thus the curtain as described has exactly the same effect as installing double-glazing.

A curtained double-glazed window has a total thermal resistance of $\sim$2 m^2 K W^{-1}, a U-value of 0.5 W m^{-2} K^{-1}, and the total rate of heat loss for the glazing under consideration for the conditions described becomes 0.84 kW.

8.4.7.4   CAVITY FLOORING
The annual mean rate of heat loss through the base of the building considered has been calculated as 0.17 kW. A 25 mm wooden floor raised on joists above the base produces a horizontal air gap which, despite not reducing overall heat losses substantially, raises the temperature of the internal surface of the flooring, so avoiding discomfiture. The additional thermal resistances introduced by the air gap and the bulk resistance of the wooden floor are $\sim$1.3 m^2 K W^{-1} and 0.18 m^2 K W^{-1} respectively, resulting in an overall thermal resistance of 3.86 m^2 K W^{-1}, a U-value of 0.26 W m^{-2} K^{-1}, and reduces heat losses through the base from 0.17 kW to 0.1 kW.

8.4.7.5 CARPETS

A 10 mm thick carpet has a thermal resistance of ~ 0.23 m^2 K W^{-1} [41]. A carpeted solid floor has therefore a total thermal resistance of 2.61 m^2 K W^{-1} and a U-value of 0.38 W m^{-2} K^{-1}, whilst a carpeted wooden raised floor has a total thermal resistance of 4.09 m^2 K W^{-1} and an overall thermal conductance of 0.24 W m^{-2} K^{-1}. The use of carpets reduces heat losses from the base of the building under consideration from 0.17 kW to 0.15 kW, or to 0.1 kW if a suspended wooden floor is also included.

8.4.7.6 ROOF INSULATION

The flat roof of the building considered experiences an annual mean rate of heat loss of 1.25 kW over 100 m^2. An additional 10 mm of wallboard underslung at the ceiling creating a 20 mm air gap would provide an extra 0.8 m^2 K W^{-1} thermal resistance. The resulting overall thermal resistance then becomes 1.92 m^2 K W^{-1}, the overall heat transfer coefficient U 0.52 W m^{-2} K^{-1}, and the rate of heat loss 0.73 kW.

A pitched roof provides an extra resistance to heat flow and also offers further scope for insulating. For example, a 150 mm layer of glass fibre laid between the joists in an attic has a thermal resistance of ~ 3.75 m^2 K W^{-1}. The introduction of a pitched roof with suitable insulation can increase the overall thermal resistance to ~ 7 m^2 K W^{-1} and so reduce heat losses through the roof of the structure considered from 1.25 kW to 0.2 kW.

8.4.7.7 SUMMARY DATA

The effects of insulating the structure described in Chapter 5 are summarised in Table 8.2. Two degrees of insulation are considered.

The greatest savings ($\sim 30\%$ of the original heat loss) for this particular building can be achieved by double-glazing or by reducing heat losses from the windows at night using thick curtains. The two levels of insulation produce energy savings of 43% and 64% respectively. Orders of priority and cost effectivenesses of insulating methods may be deduced by performing the analysis demonstrated. It is easily seen that if every existing building in the United Kingdom was insulated effectively, the nation's fuel bill could be reduced according to a total energy saving of $\sim 1.2 \times 10^{12}$ MJ per annum.

8.5 THERMAL DESIGN

The purpose of insulating systems is to form thermal barriers between environments maintained under different conditions. Hence each of these environments enter into a greater degree of thermal isolation from its neighbours. Not only are free flows of thermal energy impeded but movements of air and water across system boundaries are also inhibited. This can lead to ventilation deficiencies and problems associated with water vapour migration. These may often be overcome only by installing forced ventilating or air-conditioning equipment. The thermal design thus becomes more costly to implement.

The isolated system offers greater possibilities for control. If the intended purpose of the building is clearly defined, the method of heating or cooling may be selected rationally and the proper controls adopted. An enclosure intended for intermittent occupation (i.e. schools, shops, cinemas, and some private dwellings) will require only intermittent heating. Such a system should comprise an insulated structure with a small thermal mass. Insulation should be applied to the inside surfaces of each external wall so that the time constant for the inhabited space is reduced appropriately. Continuous occupation requires the continuous

TABLE 8.2. BENEFITS OF INSULATING

Component (area (m²))	Details of insulation	Thermal resistance R (m² K W⁻¹)	Heat transfer coefficient U (W m⁻² K⁻¹)	Heat saved[a] (kW)
	(a) First level of insulation			
Base (100)	Carpeted	2.61	0.38	0.02
Walls (280)	Insulant fill	2.50	0.40	0.32
Glazing (120)	Double-glazing or thick curtains	1.22	0.82	1.98
Roof (100)	Underslung cavity	1.92	0.52	0.52
Effective total values		2.20	0.45	2.84
	(b) Second level of insulation			
Base	Carpeted suspended floor	4.09	0.24	0.07
Walls	With convection and radiation totally suppressed	3.06	0.33	0.63
Glazing	Double glazing and thick curtains	2.00	0.50	2.52
Roof	Pitched roof with insulation laid between the joists in the attic and with underslung cavities at the roof and ceiling	7.00	0.14	1.05
Effective total values		3.51	0.28	4.27

[a] The values for the heat saved were obtained by comparison with figures presented in Table 5.3.

heating of a large thermal mass. This may often be aided by insulating each boundary at the outside surface.

Because of structural time lags, controls must predict future temperature–time responses. Thus an appropriate control system would monitor internal and external environmental variables and would adjust heater or cooler dissipations in advance in accordance with the predicted transient and periodic response of the system. Any variance from this strict form of control would lead to periods of deficient or surplus heating or cooling and hence energy wastage.

The purpose of any space-conditioning system is to maintain the internal environment at a steady temperature which is constant within acceptable limits for human thermal comfort. Heat supplied to the structural mass does not immediately fulfil this requirement. Much improved efficiency can be achieved by ensuring that the bulk of the conditioned air comes into direct contact with personnel and is not allowed to dwell in unoccupied zones (i.e. at ceilings, in roof spaces, or in empty rooms). Thus the inlet grills and diffusers of conditioned air systems should be designed to promote good air circulation, radiators should be located under windows to combat down-draughts, and fans and jets [63] be usefully employed to redirect air flows.

Chapter 9

Secondary Effects

9.1 FOOLS RUSH IN

Since the so-called "energy crisis" in 1974, domestic consumers and industrialists have been exhorted by the Government to save energy without really being told how energy is being wasted or where the priorities lie. Thus most attempts to obey this dictum have relied upon low-cost instant-return applications. Too much insulation can, in certain circumstances, lead to an increase in overall energy consumption by introducing the need to cool overheated interiors during warmer periods. Any improvement in thermal isolation restrains the free and natural passage of water vapour. The indiscriminate use of isolating techniques can create problems of incorrect ventilation and rife condensation. To avoid these difficulties the isolated structure should contain a controlled air-conditioning system.

9.2 BALANCE POINT

The analysis of thermal loads adopted in previous chapters is based upon total annual data. The capacities of winter heating and summer cooling equipment should be chosen on the basis of near-extreme climatic situations. Within these extremes, controls are required to accommodate intermediate variations. Thus instantaneous thermal loads should be estimated over the yearly cycle. As a first approximation the "load diagram" can be constructed by connecting summer and winter thermal loads (Fig. 9.1). The point where the "load line" intersects the outside air temperature abscissa indicates the "balance point" where no internal heating or cooling is required. The outside air temperature at the balance point is the "changeover" temperature for the plant (i.e. when heating equipment is substituted for cooling equipment and vice versa). The effect of insulating the system is to reduce transmission losses, to move the balance point to the left, and so reduce the changeover temperature. The total amount of heating required, represented by the area enclosed by the positive section of the load line, is reduced, whilst the total cooling load is increased. Although the capital and maintenance costs of cooling equipment are substantially higher than heating appliances, the running costs of a cooling system can be lower. The degree of insulation provided and the intended use and life of the building affects the choice of the plant. Care must be taken to avoid an increase in overall running costs resulting from the application of insulation. A degree of variable insulation is advisable (even if this merely means the ability to open windows) and in some cases it may pay to remove insulation from the system, at least during part of the year. Insulated structures designed to collect and store solar energy

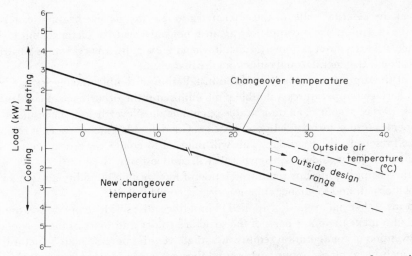

Fig. 9.1. A typical simple load diagram for a building showing the effects of insulating: ——— for an uninsulated building; ——— for an insulated building; —————— outside design range.

may run the risk of becoming fire hazards† unless proper precautions are taken at the design stage. Accurate behavioural predictions are necessary to locate these problems and to develop solutions.

9.3 VAPOUR MIGRATION AND CONDENSATION

Rates of water vapour transport between two psychrometric environments through a partition depend upon (a) the partial vapour pressure difference across the partition, and (b) the resistance of the partition to vapour migration.

The partial pressure of an air–water vapour mixture is calculated from the product of the relative humidity ϕ and the saturation vapour pressure p_{sat} at the prevailing dry-bulb temperature. Thus the higher the air temperature and the relative humidity, the greater the vapour pressure. Steam or water addition to an air environment at constant atmospheric pressure and at a constant temperature increases the specific humidity γ of the environment and hence the relative humidity. Consultations with psychrometric charts show that cold external winter environments, despite having high relative humidities, have low specific humidities and vapour pressures relative to internal conditions. Thus, in winter, positive partial pressure differences existing across the walls of buildings cause water vapour migration from inside to outside. Wherever a local psychrometric mixture is cooled below its dew-point, condensation occurs.

A general reduction in room air temperatures under conditions of constant specific humidity increases relative humidities and decreases saturation pressures but leaves the vapour pressure invariant. Reduced inside surface temperatures increase the prospect of the occurrence of condensation on the inside surfaces of external walls. Partial zonal air temperature reductions (i.e. in unoccupied rooms, bathrooms, attics, cellars, etc.) can cause condensation within, or at the surfaces of, internal walls forming boundaries with these zones,

†Many insulating materials such as expanded polystyrene and fibreboard burn very easily; polystyrene produces poisonous gases when it burns.

and/or at the external walls of the unheated rooms. Any partial heating exercise should isolate the closed-off areas completely against, not only heat transference, but also vapour movements. This is often extremely difficult when lightweight, low-resistance partitions have been installed. In general, zonal reductions are inadvisable.

One of the arguments in favour of the installation of double glazing is that by raising window surface temperatures a double pane eliminates the occurrence of condensation on the glass. This is, of course, an undesirable phenomenon. Nevertheless, the insulating process does provide a crude form of dehumidification. Good double-glazing does not remove excess moisture present in the room air. This will migrate to colder regions inside the structure (uninsulated walls or colder rooms) where it will condense out.

Weather-stripping inhibits the introduction of low specific humidity outside air and the egress of moist air and so further enhances risks of condensation.

All forms of thermal insulation applied to buildings raise inside temperatures and humidities and also make the outer parts of the structure colder and more prone to condensation. The prevention of condensation requires adequate ventilation, adequate structural heating, and the incorporation of vapour barriers and thermal insulation where necessary.

9.3.1   Effects of moisture on thermally insulated systems

When an insulant absorbs moisture its thermal conductivity rises significantly (Table 9.1). The moisture contained can also increase rates of corrosion, mould growth, and deterioration of the structural materials. If the water freezes, rupture of the insulant and its neighbouring components can occur. Radiation shielding can be rendered useless by the deposition of very thin liquid films on the surfaces because the emissivity of most liquids are very high. Condensation conditions can only be avoided by modifying constructional details (i.e. by providing an internal vapour barrier on the warmer side of the structure or by increasing the permeance of outer regions by, for example, ventilating cavities or roof spaces) or by varying the internal environmental conditions. Vapour barriers cannot be completely relied upon in the long term when continuous high humidities are present. The only safe and practical solution is to prevent condensation by proper thermal design or by air conditioning the structure.

9.3.2   Vapour diffusion

Vapour diffusion occurs through a vapour porous material whenever a difference in vapour pressure Δp_v exists across the material. Then heat losses associated with mass movements can take place even under isothermal conditions.

TABLE 9.1. EFFECTS OF MOISTURE CONTENTS ON THE THERMAL CONDUCTIVITIES OF BUILDING
MATERIALS AND THERMAL INSULANTS [19]

Material	Thermal conductivity (W m^{-1} K^{-1})	
	Dry sample at 20°C	After prolonged exposure to a wet environment
Asbestos insulating board	0.15	0.21
Asbestos slabs	0.05	0.15
Cellular concrete	0.18	0.25
Cork slab	0.04	0.05
Fibreboard	0.05	0.08
Plasterboard	0.16	0.19
Plywood	0.14	0.14

The rate of vapour flow through a material $\dot{G}_v$ $(\text{kg m}^{-2}\text{s}^{-1})$ is given by

$$\dot{G}_v = \Delta p_v \frac{k^*}{\delta},\tag{9.1}$$

where p_v is the vapour pressure (N m^{-2}), δ is the thickness of the material (m) across which the vapour pressure difference exists, and k^* is the vapour permeability of the material $(\text{kg m N}^{-1}\text{s}^{-1})$, which is a parameter analogous to thermal conductivity

The quantity k^*/δ is called the permeance C^* $(\text{kg N}^{-1}\text{s}^{-1})$ of a slab of thickness δ, and is hence analogous to thermal conductance. The permeance is used to rate the vapour diffusion characteristics of films and surface coatings rather than solid slabs. The permeability of a material is often expressed as a fraction of the permeability of air, i.e.

$$k^* = k_a^*/\kappa = \frac{1}{\kappa}\frac{D^*}{\mathbf{R}_v T},\tag{9.2}$$

where κ is a dimensionless diffusion resistance factor; D^* is the diffusion coefficient for water vapour in air, $\simeq 2.8 \times 10^{-5}\text{ m}^2\text{ s}^{-1}$; and k_a^* is the permeability of air, $\simeq 20 \times 10^{-11}\text{ kg m}$ $\text{N}^{-1}\text{ s}^{-1}$.

Thus eqn. (9.1) can be expressed alternatively as

$$\dot{G}_v = \frac{\Delta p_v}{\kappa}\frac{k_a^*}{\delta}.\tag{9.3}$$

Vapour pressure gradients may be calculated exactly as the temperature gradients through structures (i.e. composite plane walls, cylindrical structures, etc.) if the analogy between the systems of mass transfer and heat transfer equations is appreciated. Typical values for the diffusion resistance factor κ and the permeance C^* are given in Tables 9.2 and 9.3. It must be noted that these values vary extensively between samples because of the heterogeneous nature of the substances involved and because other variable factors (i.e. material compression, method of application, moisture content, arrangement of constituents, etc.) also

TABLE 9.2. DIFFUSION RESISTANCE FACTORS [19]

Material	Diffusion resistance (factor κ)
Brick	10
Brickwork	35
Cement mortar	45
Clinker block	420
Concrete	40
Cork slab	10
Foam glass	zero
Hardboard	150
Insulating board	5
Mineral wool	1
Plaster	10
Plasterboard	8
Plywood	200–700
Woods	20–300
Air	1

TABLE 9.3. PERMEANCE

Material	Permeance C^* $(10^{-11}$ kg N^{-1} s$^{-1})$ [19]
Aluminium foil	0–0.6
Building paper	1.7–45.0
Painted insulating board	5.7–290.0
Kraft paper	170–460
Painted plaster	5.7–17.0
Painted plywood	14.0
Polythene	0.6
Roofing felt	1.0–23.0
Painted wood	1.7–11.0
Aluminium painted wood	25.0–54.0

substantially affect the values obtained by standard measurements. Condensation sites within semi-solid systems may be located by:

(a) computing temperature distributions in the usual way;
(b) computing vapour pressure distributions in an analogous way;
(c) superimposing saturation curves corresponding to the local temperatures.

Condensation will occur wherever the calculated vapour pressure is equal to the local saturation pressure. Whereas this analysis represents a steady-state procedure, transient behaviour should also be studied to avoid instantaneous condensation during cycling or transient adjustment to changes in heating or cooling provisions.

For water vapour in air, a surface coefficient for mass transfer is related to the heat transfer coefficient via the Lewis relationship

$$h^* = h_c / \rho c_p,\tag{9.4}$$

where h^* is a surface mass transfer coefficient (m s^{-1}). At ordinary room temperatures, the air density $\rho \simeq 1.2$ kg m^{-3} and the specific heat $c_p \simeq 1000$ J kg^{-1} K^{-1}, and so

$$h^* \simeq h_c / 1200.\tag{9.5}$$

The total amount of mass transferred $\dot{G}_v$ (kg m^{-2} s^{-1}) is given by

$$\dot{G}_v = \frac{h^*}{\mathbf{R}_v T}(p_v - p_{sat}),\tag{9.6}$$

*, a negative value of $\dot{G}_v$ indicates that evaporation is occurring.
Where $\mathbf{R}_v$ is the characteristic gas constant for water vapour $= 461$ J kg^{-1} K^{-1}, T is the mean absolute temperature (K), p_v is the vapour pressure in the bulk fluid (N m^{-2}), and p_{sat} is the saturation vapour pressure of the fluid at the surface temperature (N m^{-2}).
At 20°C, $\mathbf{R}_v T \simeq 135\,000$ J kg^{-1} and hence

$$\dot{G}_v = \frac{h_c}{1.62 \times 10^8}(p_v - p_{sat}).\tag{9.7}$$

Any suitable expression listed in Chapter 2 may be used for the estimation of the convective heat transfer coefficient h_c.

Vapour resistances $R_v{}^*$

(i) For an air layer,

$$\text{vapour resistance } R_v^* = \frac{\mathbf{R}_v T}{h^*} \text{ N s kg}^{-1}$$

but

$$h^* = h_c/\rho c_p \text{ m s}^{-1}$$

$$= h_c/1200,$$

and

$$\mathbf{R}_v = 461 \text{ J kg}^{-1} \text{ K}^{-1}.$$

Thus

$$\underline{R_v^* = 1.62 \times 10^8 \, R_c \text{ N s kg}^{-1}.}$$

(ii) For a solid member,

$$\text{vapour resistance } R_v^* = \frac{\delta}{k^*} = \frac{\delta\kappa}{k_a^*}$$

and

$$k_a^* = 20 \times 10^{-11} \text{ kg m N}^{-1} \text{ s}^{-1}.$$

Thus

$$R^* = 5 \times 10^9 \, \delta\kappa \text{ N s kg}^{-1}.$$

(iii) For a surface coating or film,

$$\text{vapour resistance } R_v^* = (C^*)^{-1} \text{ N s kg}^{-1}.$$

9.3.3 Condensation

Condensation of water occurs whenever moist air comes into contact with a surface whose temperature is below the prevailing dew-point temperature of the psychrometric mixture. Condensation cannot take place if the surface is at a temperature above the dew-point. Condensation or evaporation is accompanied by the release or absorption of latent heat H_{fg} (J kg^{-1}) being the latent heat transferred per kg during the phase change. For water at 20°C, $H_{fg} \simeq 2.5 \times 10^6$ J kg^{-1}, and so the total rate of heat dissipation or absorption is calculated from

$$\dot{Q} = \dot{G}_v H_{fg} \text{ (W m}^{-2}). \tag{9.8}$$

9.3.4 Examples of application

(a) *Unventilated air cavity.* Two 100 mm layers of brickwork are separated by a 50 mm thick air layer. Inside and outside air conditions are $T_i = 25°C$, $\phi = 80\%$, and $T_o = 11°C$ with $\phi = 100\%$ respectively. Using the numbering sequence for nodes adopted in Chapter 5.

Node	i	1	2	3	4	0	Units
Temperature T	25	22.5	20.7	15.3	13.5	11.0	°C
Thermal resistance R		0.48	0.33	1.02	0.33	0.48	m² K W⁻¹
Saturation vapour pressure p_s (Fig. 2.14)	3.17	2.6	2.45	1.75	1.57	1.33	10³ N m⁻²
Relative humidity ϕ	80					100	%
Vapour resistance R_v^*		0.77	175	1.65	175	0.77	10⁸ N s kg⁻¹
Vapour pressure p_v	2.54	2.54	1.94	1.93	1.33	1.33	10³ N m⁻²

Total thermal resistance $R_{\text{tot}} = 2.64$ m² K W⁻¹

Total vapour resistance $R_{v,\text{tot}}^* = 353 \times 10^8$ N s kg⁻¹

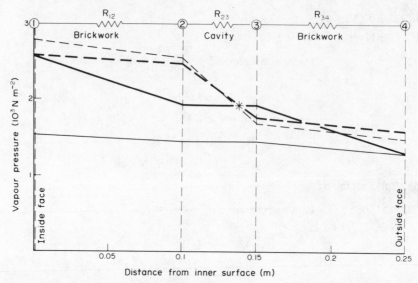

Fig. 9.2. Water vapour transmission and condensation within a wall: —— vapour pressure distribution within a wall containing an unventilated air cavity or with an inter-cavity insulant present (mineral wool); — — — saturated vapour pressure distribution for the wall with an unventilated air cavity; ———— saturated vapour pressure distribution for the wall filled with insulant with or without a vapour barrier applied to the inside face; —— vapour pressure distribution for the insulated wall with a vapour barrier applied to the inside face; * condensation site.

Figure 9.2 shows that condensation will occur at the inner face and inner section of the outer layer of brickwork. This may be avoided by venting the cavity (and so reducing the external resistance to vapour flow) or by introducing a vapour barrier on the warmer side of the system (and so increasing the internal resistance to vapour flow).

(b) *Mineral-wool-insulated cavity*

Node	i	1	2	3	4	0	Units
Temperature T	25	22.8	21.4	14.6	12.2	11	°C
Thermal resistance R		0.48	0.33	1.50	0.33	0.48	m^2 K W^{-1}
Saturation vapour pressure p_s	3.17	2.75	2.52	1.67	1.47	1.33	10^3 N m^{-2}
Relative humidity ϕ	80					100	%
Vapour resistance R_v^*		0.77	175	5	175	0.77	10^8 N s kg^{-1}
Vapour pressure p_v	2.54	2.54	1.94	1.93	1.33	1.33	10^3 N m^{-2}

Total thermal resistance $R_{tot} = 3.12$ m^2 K W^{-1}
Total vapour resistance $R_{v,tot}^* = 365.5 \times 10^8$ N s kg^{-1}

The introduction of mineral wool increases the temperatures, and hence the saturation pressures, on the warmer side of the structure, whilst depressing temperatures and saturation pressures on the colder side (Fig. 9.2). Because the resistance of the mineral wool contributes little to the overall vapour resistance, the vapour pressure distribution is not significantly affected. Condensation within the insulant will occur and the risk of condensation in the outer layer is increased. Venting-off under cyclic conditions is somewhat inhibited by the presence of the insulant.

(c) *Insulated cavity with a vapour barrier.* The introduction of a sheet of foil-backed building paper $C^* = 0.6 \times 10^{-11}$ kg N^{-1} s^{-1}) at the inner surface of the inner wall does not affect the saturation vapour pressure distribution but does lower local vapour pressures.

Node	i	1	2	3	4	o	Units
Temperature T	25	22.8	21.4	14.6	12.2	11	°C
Thermal resistance R		0.48	0.33	1.50	0.33	0.48		m^2 K W^{-1}
Saturation vapour pressure p_s	3.17	2.75	2.52	1.67	1.47	1.33	10^3 N m^{-2}
Relative humidity ϕ	80					100	%
Vapour resistance R_v^*		1666	175	5	175	0.77		10^8 N s kg^{-1}
Vapour pressure p_v	2.54	1.54	1.43	1.43	1.33	1.33	10^3 N m^{-2}

Vapour resistance of the vapour barrier $R_v^* = (C^*)^{-1} = 1.66 \times 10^{11}$ N s kg^{-1}
Total vapour resistance $R_{v,\text{tot}}^* = 2022 \times 10^8$ N s kg^{-1}

It may be seen (Fig. 9.2) that the vapour pressure is lower than the saturation vapour pressure at all positions in the structure, thus internal condensation will be avoided. Alternative means to remove latent heat from the occupied space must, however, be provided.

9.4 VARIATION OF THE THERMAL CONDUCTIVITY OF AN INSULANT WITH MOISTURE CONTENT

Once condensation has taken place within an insulant (or building material), its effective thermal and mass transport properties change. This modifies the temperature and vapour pressure distributions through the system and further calculations are required to obtain an indication of quasi-steady-state behaviour. Unfortunately, there is a lack of reliable information quantifying the effects of moisture content on the overall transport properties of common materials. Most thermal conductivity values for wet insulants have been obtained from uncorrected measurements using crude conductivity probes in saturated materials, or with uniform moisture content. Neither has the degree of moisture penetration been adequately specified (see Table 9.1). A major research effort is obviously required to examine systematically the quasi-steady-state and transient and periodic responses of damp-insulated systems.

A first-order indication of the effects of water content in porous materials can be obtained by assuming that the thermal resistance of an insulant is dependent upon the amount of dry air contained. The overall conductivity of the dry insulant then depends upon the volume voidage of the medium and the conductivity of the solid constituent, i.e.

$$k_{\text{dry}} = V_{\text{air}}k_a + (1 - V_a)k_b. \tag{9.9}$$

Because any liquid water contained in damp insulants displaces its own volume of air,

$$k_{\text{wet}} = V_a k_a + V_{\text{water}}k_{\text{water}} + (1 - V_a - V_{\text{water}})k_b, \tag{9.10}$$

where V_a and V_{water} are the volume voidages of the air and water respectively, and k_a, k_{water}, and k_b are the thermal conductivities of the air, water, and solid component of the insulant.

EXAMPLE

The thermal conductivity of dry uncompressed mineral wool is ~ 0.04 W m^{-2} K^{-1}, and air has a thermal conductivity of ~ 0.025 W m^{-2} K^{-1}. The thermal conductivity of solid mineral wool, obtained by compressive testing, approaches 0.25 W m^{-2} K^{-1}. Thus the only unknown in eqn. (9.9) is the volume fraction of the air contained V_a, and so

$$V_a = \frac{0.04 - 0.25}{0.025 - 0.25} = 0.933.$$

This value could, of course, be alternatively obtained by direct observation of area voidages under the microscope.

Assuming that water penetrates only into the interstices, for any moisture content,

$$V_a + V_{\text{water}} = 0.933.$$

The thermal conductivity of water at 20°C is ~ 0.57 W m^{-2} K^{-1}, and so eqn. (9.10) becomes

$$k_{\text{wet}} = (0.933 - V_{\text{water}})0.025 + V_{\text{water}}0.57 + 0.067 \times 0.25$$

$$= 0.039 + 0.545\, V_{\text{water}}.$$

This expression is comparable with a characteristic for wet mineral wool obtained using a thermal conductivity probe (Fig. 9.3). It may be seen that the approximate procedure gives a linear relationship which predicts conductivity values within the limits of experimental measuring errors.

Figure 9.3 presents experimentally derived relationships for various insulants. The non-linearity of these characteristics indicates transient and stratification effects.

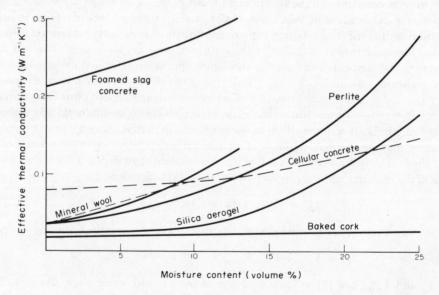

Fig. 9.3. Effects of moisture content upon the effective thermal conductivities of various insulants: — — — calculated estimate for mineral wool.

9.5 **INTEGRATED STUDIES**

Developments in energy costing may show that the energy used in the manufacture of the components of energy-saving systems could significantly offset the expected overall savings. Design for energy conservation is not an arbitrary procedure. Cash-based factors influencing the choice of energy-conserving systems favour the adoption of simple low-cost methods. Consideration is seldom given to the secondary undesirable effects arising from the application of these methods or to more-expensive but in the long term more profitable, alternatives. The well-balanced systematic design considers, in advance of implementation, all technical possibilities for systems and applications and the interrelations of all primary and secondary aspects of energy-saving techniques.

Chapter 10

Waste Heat Recovery

10.1 HEAT REJECTION AND RECLAIM

The degradation of energy from its stable form, locked up in the chemical structures of fossil fuels, to entropic disarray takes place during a combination of the following fundamental processes:

(a) energy release via combustion in which chemical energy is converted to thermal energy;

(b) energy transformation to alternate forms (i.e. to mechanical or electrical energy);

(c) energy utilisation;

(d) energy rejection to the environment.

Partial energy rejection in the form of heat, light, noise, or vibration occurs during each process and eventually all energy released is downgraded and dumped on the environment to be dissipated to outer space. Reject heat can be recovered at almost any station along the energy chain and redirected for a lower grade useful purpose, such as space conditioning, air preheating in prime movers, drying, or desalination. Waste heat recovery improves the overall efficiency of energy utilisation.

Each reclaim system employs some form of heat exchanger of which there are two basic classes: recuperators and regenerators. Whereas a recuperative heat exchanger allows heat to be transferred in a steady-state process between two fluids at different temperatures separated by a wall, a regenerative heat exchanger employs a secondary heat transfer medium which charges and discharges periodically during a cyclic transient process.

10.2 RECUPERATIVE HEAT EXCHANGERS

In the recuperator, heat is transferred between two flowing fluids across a solid wall (Fig. 10.1). The rate of heat transfer over an area A is calculated from

$$\dot{Q} = UA\Delta T, \tag{10.1}$$

where U (W m^{-2} K^{-1}) is the overall heat transfer coefficient connecting the bulk fluid temperatures either side of the wall and ΔT (K) is a mean difference between these temperatures over the region considered. The overall heat transfer coefficient is calculated by adding

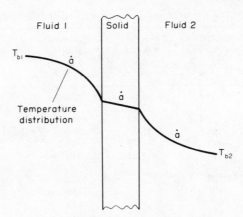

Fig. 10.1. Recuperative heat transfer.

the reciprocals of the two partial fluid convective heat transfer coefficients to the resistance of the solid wall as follows:

$$\frac{1}{UA} = \frac{1}{h_1 A} + \frac{\delta}{k_s A} + \frac{1}{h_2 A}.$$ (10.2)

Although the area A cancels out in eqn. (10.2), it has been left in the relationship purposefully because heat transfer areas vary through radial flow systems. Fluids are usually arranged to flow turbulently through heat exchangers to obtain the high rates of heat transfer associated with turbulent flow. Mean surface to fluid Nusselt numbers may be obtained from empirically or otherwise derived equations of the form

$$\overline{Nu} = \text{constant} \times Re^n Pr^{0.33},$$ (10.3)

where the constant depends upon the configuration of the heat transfer surface and the index n generally lies between 0.6 and 0.8. The overall heat transfer coefficient is usually assumed to be uniform, and its value is often calculated at a representative position half-way through the flow passages of the heat exchanger.

10.2.1   Typical values of individual and overall heat transfer coefficients

The values given in Table 10.1 are indicative of the order of magnitude of the heat transfer coefficients to be expected for clean heat transfer surfaces. The values are commonly found in practice and may be predicted from the turbulent flow equation

$$\overline{Nu} = 0.023 \, Re^{0.8} Pr^{0.33}.$$ (2.49)

Because a heat exchanger's function is to enhance the transference of heat, filmside heat transfer coefficients should be as large as possible. This may be achieved by employing relatively fast flow rates, and so quite high pumping powers are necessary. It is usual to compromise on values of overall heat transfer coefficients and pressure drops by arranging that $Re \sim 50\,000$. The velocities listed in Table 10.1 are therefore typical values compatible with pressure drop considerations. The values apply for parallel or counterflow concentric tube heat exchangers with a steel tube wall of thickness $\sim 3\,\text{mm}$ ($k = 45\,\text{W m}^{-2}\,\text{K}^{-1}$).

TABLE 10.1. TYPICAL VALUES OF INDIVIDUAL AND OVERALL HEAT TRANSFER COEFFICIENTS

Hot side to cold side	Partial hot side coefficient h_{hs} (W m^{-2} K^{-1})	Hot side velocity u_{hs} (m s^{-1})	Partial cold side coefficient h_{cs} (W m^{-3} K^{-1})	Cold side velocity u_{cs} (m s^{-1})	Overall heat transfer coefficient U (W m^{-2} K^{-1})
Air to air	28.4	4.58	28.4	4.58	14.2
Air to air	56.8	12.2	56.8	12.2	28.4
Air to water	28.4	4.58	4090	1.52	28.1
Air to water	56.8	12.2	4090	1.52	56.2
Water to water	4090	1.52	4090	1.52	1760
Oil to oil	510	1.52	510	1.52	256
Condensing steam to boiling water	11 400	—	5768	—	2980
Condensing steam so water	11 400	—	1136	1.52	960
Condensing steam to oil	11 400	—	510	1.52	477
Oil to water	510	1.52	4090	1.52	440
Oil to air	510	1.52	28.4	4.58	25.8

Equation (10.2) illustrates the process of adding resistances to heat flow as the heat transfers from a hot fluid to a colder fluid. It is evident from the values given in Table 10.1 that the resistance of the tube wall ($\delta/k = 1/14\,250$) offers little heat transfer resistance compared with the partial filmside resistances. Thus in many calculations the resistance of the tube wall may be neglected and so the same value of the overall heat transfer coefficient appertains irrespective of the material used for the heat exchanger construction. It may be deduced that, whilst it seems intuitive that a boundary designed to allow the free passage of heat should be fabricated from a material with a high thermal conductivity (such as copper and brass), this is not strictly the case, and any solid material will do as long as its structural and chemical characteristics are compatible with non-heat transfer requirements. The only exception to this rule is for extended surface heat exchangers where, in order to achieve uniformity of temperature over the fins, and thus a high fin effectiveness, good thermal contact should exist between the secondary and primary surfaces, and the extended surface should have a high thermal conductivity.

10.2.2 Secondary surfaces

Equation (10.2) has the characteristic that the overall heat transfer coefficient is always less than the smaller partial heat transfer coefficient contributing to it. Thus when one partial coefficient is appreciably smaller than the other (such as in the situation where heat is being transferred between air and water), the overall heat transfer coefficient—being controlled by the smaller value—is little influenced by altering the magnitude of the larger partial coefficient. In such circumstances it often becomes economical to fit more heat transfer area to the surface in contact with the fluid having the lower partial heat transfer coefficient by the addition of a secondary surface on that side. This is achieved by joining high-conductivity fins (spirally wound or "gills") to the primary surface which separates the fluids. By these means the lower partial heat transfer coefficient can be effectively increased by the ratio of the areas of the surfaces in contact with the two fluids. Then eqn. (10.2) becomes (ignoring metal conductivity and dirt resistance)

$$\frac{1}{U} = \frac{\xi A_1}{h_2 A_2} + \frac{1}{h_1},$$
(10.4)

where ξ is a fin-effectiveness factor $(0-1)$ which describes the efficiency of the extended surface in maintaining temperature uniformity over the extended surface of area A_2 [64]. It is important to observe that the value for the overall heat transfer coefficient U is usually referred to the smaller surface area.

10.2.3 Mean temperature difference

So far only an element of the heat transfer surface has been considered. Throughout a heat exchanger, the temperature of each fluid (unless condensation or evaporation is taking place) is continuously varying, the hotter fluid being cooled and the colder fluid being heated. For concentric-flow heat exchangers it can be shown that the mean temperature difference $\overline{\Delta T}$ that satisfies eqn. (10.1) is the logarithmic mean of the temperature differences at the two ends of the heat exchanger [18].

Heat exchanger operations may be classified into three groups:

(1) *Parallel flow operation* (Fig. 10.2a) in which the two fluids flow in the same direction along parallel paths separated by a wall.
(2) *Counter-flow or contra-flow operation* (Fig. 10.2b) in which the two fluids flow in opposite directions along parallel paths separated by a wall.
(3) *Cross-flow operation* (Fig. 10.2c) in which the two fluids flow at right angles to one another separated by a wall.

The logarithmic mean temperature difference $\overline{\Delta T}_o$ to be employed whilst estimating heat transference across parallel and counter-flow heat exchangers is obtained [18] from

$$\overline{\Delta T}_o = \frac{\text{largest terminal temperature difference} - \text{smallest terminal temperature difference.}}{\log_e \dfrac{\text{largest terminal temperature difference}}{\text{smallest terminal temperature difference}}} \quad (10.5)$$

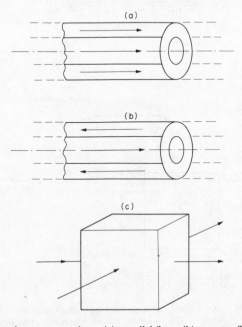

Fig. 10.2. Heat exchanger operations: (a) parallel flow; (b) counter-flow; (c) cross-flow.

10.2.4   **Types of recuperative heat exchangers**

Practical recuperators are far more complex than single concentric pipe arrangements. Oil coolers and large marine heat exchangers are often of shell and tube construction in which one fluid passes through a straight bundle of parallel tubes whilst the other fluid flows over these tubes (Fig. 10.3a). Suitable inlet and exit ports direct the fluids to achieve single or multiple pass flows on either side of the heat exchanger, and baffles bring the fluids into good thermal contact with the heat transfer surfaces by redirecting and accelerating local flows. Vehicle air-cooled "radiators" are usually cross-flow devices (Fig. 10.3b) with extended surfaces on the air side. Because these systems cannot be simply classified as either parallel or counter-flow arrangements, suitable correction factors F^* (taking values between zero and unity) are available from analytical or, more usually, empirical data, which can be applied to imaginary counter-flow logarithmic mean temperature differences [18, 21, 64].

10.2.5   **The design and selection of recuperators**

When an intended heat exchange system is being designed, all the terminal temperatures are usually prescribed, and so the heat transfer area required may be determined from analyses involving mean temperature differences. Then

the heat lost by the hot side fluid = the heat transferred = the heat gained by the cold side fluid.

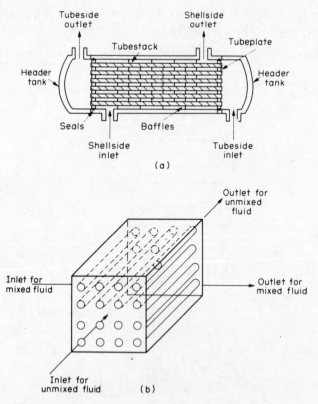

Fig. 10.3. Sketches of practical recuperators: (a) a counter-flow single-pass shell and tube heat exchanger; (b) a cross-flow compact heat exchanger with one fluid mixed, the other unmixed, during flow through the device.

Using the subscripts cs, hs, I and O to indicate cold side, hot side, and inlet and outlet conditions,

$$Q = (\dot{m}c_p \Delta T)_{hs} = UA\overline{\Delta T}F^* = (\dot{m}c_p \Delta T)_{cs}, \qquad (10.6)$$

where $\dot{m}$ represents mass flow rates (kg s^{-1}) and c_p (J kg^{-1} K^{-1}) mean specific heats.

$$T_{cs} = T_{csO} - T_{csI}, \quad T_{hs} = T_{hsI} - T_{hsO},$$

and

$$\overline{\Delta T} = \frac{(T_{hsI} - T_{csO}) - (T_{hsO} - T_{csI})}{\log_e \dfrac{(T_{hsI} - T_{csO})}{(T_{hsO} - T_{csI})}} \cdot \dagger$$

The product $\dot{m}c_p$ will be expressed more simply as M. The mass flow rates and the areas to flow selected depend upon the permissible pressure drops compatible with pumping arrangements and energy conservation requirements.

For a pipe or duct the pressure drop Δp (Nm^{-2}) is calculated from

$$\Delta p = f \frac{L}{D_{hd}} \frac{\rho u^2}{2}, \qquad (2.15)$$

and this expression may often be applied without excessive error to more complex passageways. The pumping work required for each side of the exchanger W (W) is then estimated from

$$W = \Delta p A u, \qquad (10.7)$$

where A is the area to heat flow (m^2). In heat reclaim applications the dimensionless ratio of the rate of heat reclaimed Q to the pumping power necessary W is a useful additional parameter by which to compare different arrangements.

10.2.6 Heat exchanger effectiveness

Equation (10.1) can only be used when all the terminal temperatures are known. In many cases, however, particularly in off-design operation, whilst the overall heat transfer coefficient can be estimated with reasonable accuracy, the temperatures of the fluids leaving the exchanger are not known. The outlet temperatures can only then be calculated using the logarithmic mean temperature difference approach by a tedious "trial and error" process. Fortunately, this may be avoided by using the concept of heat exchanger effectiveness which does not involve the use of a mean temperature difference. Heat exchanger effectiveness is defined as

$$\omega = \frac{\text{actual heat transfer accomplished}}{\text{maximum heat transfer possible}}. \qquad (10.8)$$

The maximum possible rate of heat transfer would be accomplished in a heat exchanger of infinite heat transfer area. Then the temperatures at one terminal of the heat exchanger would be identical [18].

The *actual heat transfer accomplished* is equal to the heat lost by the hot fluid or that gained by the cold fluid, i.e.

$$Q = M_{hs}(T_{hsI} - T_{hsO}) = M_{cs}(T_{csO} - T_{csI}). \qquad (10.9)$$

---

†The case $M_{cs} = M_{hs}$, which is commonly encountered in heat reclaim from exhaust air, leads to an indeterminate value of ΔT via this expression when the fluids cross in counter-flow. The value of $\overline{\Delta T}$, which should then be used, is calculated from the arithmetic mean of the terminal temperature differences.

The maximum possible rate of heat transfer depends upon the ratio of the thermal mass flows of each fluid.

When $M_{hs} > M_{cs}$,

$$\frac{M_{hs}}{M_{cs}} = \frac{T_{csO} - T_{csI}}{T_{hsI} - T_{hsO}} > 1.0, \qquad (10.10)$$

but the highest value T_{csO} can reach is T_{hsI}, and so the maximum possible rate of heat transfer would occur if T_{csO} equalled T_{hsI}, then

$$\dot{Q}_{max} = M_{cs}(T_{csO} - T_{csI}) = M_{cs}(T_{hsI} - T_{csI})$$

and

$$\omega = \frac{M_{hs}(T_{hsI} - T_{hsO})}{M_{cs}(T_{hsI} - T_{csI})}. \qquad (10.11)$$

Thus if the effectiveness of a heat exchanger is known, T_{hsO} can be calculated from eqn. (10.11) and then T_{csO} can be calculated from eqn. (10.9). Alternatively, when $M_{hs} < M_{cs}$,

$$\frac{M_{hs}}{M_{cs}} = \frac{T_{csO} - T_{csI}}{T_{hsI} - T_{hsO}} < 1.0, \qquad (10.12)$$

but the lowest value T_{hsO} can reach is T_{csI}, and so the maximum possible rate of heat transfer would occur if T_{hsO} equalled T_{csI}, then

$$\dot{Q}_{max} = M_{hs}(T_{hsI} - T_{hsO}) = M_{hs}(T_{hsI} - T_{csO})$$

and

$$\omega = \frac{M_{cs}(T_{csO} - T_{csI})}{M_{hs}(T_{hsI} - T_{csI})}. \qquad (10.13)$$

Thus if the effectiveness is known, T_{csO} may be calculated from eqn. (10.13) and then T_{hsO} can be calculated from eqn. (10.9).

Equations (10.11) and (10.13) may be expressed alternatively as

$$\dot{Q} = \omega M_{min}(T_{hsI} - T_{csI}) = M_{hs}(T_{hsI} - T_{hsO}) = M_{cs}(T_{csO} - T_{csI}). \qquad (10.14)$$

Furthermore, it may be proved that [64], for a counter-flow heat exchanger,

$$\omega = \frac{1 - \exp[UA(1/M_{hs} - 1/M_{cs})]}{M_{min}\left\{\dfrac{1}{M_{cs}} - \dfrac{1}{M_{hs}}\exp[UA(1/M_{hs} - 1/M_{cs})]\right\}} \qquad (10.15)$$

and, for a parallel flow heat exchanger,

$$\omega = \frac{1 - \exp[-UA/M_{hs}(1 + M_{hs}/M_{cs})]}{M_{min}\left[\dfrac{1}{M_{hs}} + \dfrac{1}{M_{cs}}\right]}. \qquad (10.16)$$

The analytical evaluation of the effectivenesses ω for other arrangements is sometimes possible but always tedious. Usually, values for ω for various types of heat exchanger are presented in the form of graphs (see, for example, references [18] and [64] where the effectiveness is plotted in terms of the number of transfer units (NTU) and the ratio M_{max}/M_{min}). By definition

$$NTU = UA/M_{min}. \qquad (10.17)$$

10.2.7 **Example of heat exchanger assessment**

Significant reductions in the overall energy requirements of buildings can be accomplished using waste heat recovery only when the fresh-air load is an appreciable factor of the overall heat losses. The simple uninsulated building described in Chapter 5 required an energy consumption associated with ventilation of only 8% of the total annual heating load. After insulating (cf. Chapter 8), this fraction rose to 13% of the remaining energy requirement. Air-to-air heat exchangers, predominantly because of the need for extended surfaces, are notoriously expensive, and so the introduction of an air-to-air heat reclaim system would be uneconomic for this scale of application. Useful waste heat recovery from exhaust air can be obtained for larger occupied spaces having greater ventilation rates (i.e. in indoor swimming pools, restaurants, kitchens, schools, and hospitals).

In order to demonstrate the selection process, therefore, a building 30 m by 40 m by 15 m high requiring an air change rate of one per hour will be considered. Inside and outside environmental temperatures are assumed to be 25°C and 0°C respectively.

Internal volume of the building $V = 18\,000\ \mathrm{m}^3$.

Mean air transport properties

Density $\rho = 1.24\ \mathrm{kg\ m^{-3}}$.

Specific heat $c_p = 1004\ \mathrm{J\ kg^{-1}\ K^{-1}}$.

Dynamic viscosity $\mu = 1.76 \times 10^{-5}\ \mathrm{kg\ m^{-1}\ s^{-1}}$.

Thermal conductivity $k = 0.025\ \mathrm{W\ m^{-1}\ K^{-1}}$.

Prandtl number $Pr = 0.71$.

Volume flow of air $\dot{V} = 18\,000\ \mathrm{m^3\ h^{-1}} = 5\ \mathrm{m^3\ s^{-1}}$.

Mass flow of air $\dot{m} = 6.2\ \mathrm{kg\ s^{-1}}$.

$M_{max} = M_{min} = \dot{m}c_p = 6224.8\ \mathrm{W\ K^{-1}}$.

Rate of heat rejection $= mc_p\Delta T = 155\ \mathrm{kW}$.

Thus the incoming fresh air must be heated at a rate of 155 kW. Some of the heat rejected in the used exhaust air may be used to preheat the fresh air induced from the external environment.

It is assumed that an air-to-air cross-flow recuperator (Fig. 10.4) having the parameters detailed below is available.

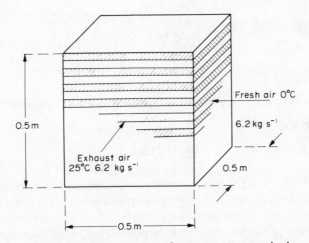

Fig. 10.4. Details of the cross-flow recuperator examined.

Details of recuperator

  Volume $= 0.125$ m^2.

  Number of air channels per side $= 25$.

  Area to flow/side $A = 0.125$ m^2.

  Air velocities $= u = \dot{V}/A = 40$ m s^{-1}.

  Wetted perimeters $L_p = 25$ m.

  Extended surface factor (corrugated surfaces) $= 5$.

  Heat transfer area $A_{ht} = 60$ m^2.

  Effective hydraulic diameter/side $D_{hd} = 4A/L_p = 0.02$ m.

Heat transfer coefficient

Reynolds numbers $= \dfrac{\rho u D_{hd}}{\mu} = \dfrac{(\dot{m}/A)D_{hd}}{\mu} = 56363$ (turbulent).

Nusselt numbers $= \overline{Nu} = 0.023\, Re^{0.8} Pr^{0.33} = 130$.

Partial heat transfer coefficients $\bar{h}_{cs} = \bar{h}_{hs} = \dfrac{\overline{Nu}\,k}{D_{hd}} = 162.5$ W m^{-2} K^{-1}.

Overall heat transfer coefficient U:

$\qquad U^{-1} = h_{cs}^{-1} + h_{hs}^{-1}$ (neglecting the resistance of the metal).

$\qquad U = 81$ W m^{-2} K^{-1}.

Number of transfer units

$$\mathrm{NTU} = \frac{AU}{M_{min}} = 0.81.$$

  Reference to effectiveness charts for a cross-flow heat exchanger [64] yields for NTU $= 0.81$ and $M_{max}/M_{min} = 1$, $\omega = 0.4$.

  Thus heat transfer accomplished $\dot{Q} = \omega M_{min}(T_{hsI} - T_{csI})$

$$= 62 \text{ kW}.$$

Power requirements

$$W = \Delta p A u$$

$$= f \frac{L}{D_{hd}} \frac{\rho u^2}{2} Au.$$

From Fig. 2.3, for smooth surfaces, $f = 0.0185$, and so $W = 2.3$ kW per side.

Heat reclaim/power ratio

$$\frac{\dot{Q}}{2W} = 13.56.$$

Percentage reclaim $= 40\%$. Thus the standard compact recuperator described may be used to preheat the incoming fresh air by $10\,^{\circ}$C. The power requirement of the system is small ($\sim 7\%$ of the heat reclaimed), and so further savings may be achieved by optimising the size and shape of the recuperator. This, however, could be possible only if the capital available is sufficient to meet the cost of non-standard components.

10.3 REGENERATIVE HEAT EXCHANGERS

In a regenerative heat exchanger the hot and cold fluids pass alternatively over the same heat transfer surface. This surface consists of one or more flow passages which are partially filled with either solid pellets, refractory solids (for high temperatures), or metal matrices (for lower temperatures). Regenerators are usually used to reclaim heat from gases or other fluids with low thermal capacities. They may be used at extremely high temperatures or for gases carrying suspended ash and solids where a conventional recuperator may become silted and blocked. During one part of the cycle, the inserts store heat in internal energy as the hot fluid passes over their surfaces. In the second part of the cycle this internal energy is released as the colder fluid passes through the regenerator. Regenerative heat transfer systems may operate in parallel flow, cross-flow, or counter-flow and may be single or multipass arrangements. The regenerator's principle advantage is that it has a high heat exchanger effectiveness per unit weight and volume; thus much more compact heat transfer surfaces can be employed relative to recuperators (i.e. a 24 mesh matrix has 1000 m^2/m^3 [64]). The heat transfer surfaces are generally substantially less expensive per unit amount of transfer area, and, because of flow reversals, no permanent fluid stagnation regions occur and systems are essentially self-cleaning. The major design problem is to prevent leakage between the fluids, especially at elevated temperatures and pressures. Regenerators have been used successfully as air preheaters in open-hearth- and blast-furnaces, in gas liquefaction plants, and in waste heat recovery processes. Latent heat transfer can also be accomplished by coating the capacity matrix with a suitable hygroscopic adsorbent (i.e. lithium chloride).

10.3.1 Types of regenerator

There are two basic types of periodic flow regenerator: rotary systems and valved systems (Fig. 10.5). Rotary systems may operate in either axial or radial flow. The valved arrangement has two identical matrices, each one functioning as either a hot flow or cold flow matrix by means of a periodic switching of the control valves. Design procedures apply in the same manner for both rotary and valved systems since the product (the mass of a single rotary matrix) × (revolutions per second) is considered in the design theory to be equivalent to the quotient (mass of the two valved matrices)/(the period of the valve operations).

10.3.2 Design of regenerators

The design theory is complex [65] evolving effectiveness–NTU relationships which require computer solution. Fortunately, values for effectivenesses have been tabulated [64] and plotted as a function of matrix capacity rate, a modified number of transfer units, NTU′, and the ratios of M_{max}/M_{min}, i.e.

$$\omega = f(\text{NTU}', M_{max}, M_{min}, M_{rot}/M_{min}), \tag{10.18}$$

where the modified number of transfer units (NTU′) is defined as

$$\text{NTU}' = \frac{1}{M_{min}}\left(\frac{1}{hA_{cs}} + \frac{1}{hA_{hs}}\right) - 1 \tag{10.19}$$

and the matrix capacity rate M_{rot} is calculated from

$$M_{rot} = (\text{mass rate}) \times (\text{specific heat of the solid})$$
$$= (\text{revolutions per second}) \times (\text{matrix mass}) \times c_{ps}. \tag{10.20}$$

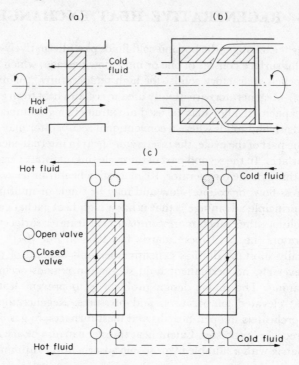

Fig. 10.5. Schematic representations of various types of regenerative heat exchangers: (a) an axial flow rotary regenerator; (b) an inward radial flow rotary regenerator; (c) valved-type periodic flow regenerative system.

Once the value of the effectiveness has been established, the design procedure is exactly similar to that for the recuperator.

10.4 RUN-AROUND COILS

The use of single recuperators and regenerators for air-to-air heat reclaim applications has the disadvantage that the fresh-air intake and the used-air exhaust must pass through the same location in the building. A run-around coil [45] is a liquid-coupled indirect-type heat exchanger [64] in which individual multirow finned heat exchanger coils are placed in the inlet and exhaust air streams. Water, pumped through small-bore insulated pipes, transfers the heat between these two air-to-water recuperators. Thus the heat exchangers can be placed conveniently far apart and plant layout remains unrestricted (Fig. 10.6).

10.5 HEAT PIPE

The transport of heat associated with vapour diffusion through a gas contributes little to the overall rate of heat transfer. However, very high heat transfer rates across small temperature gradients can be accomplished when phase changes (condensation and evaporation) occur. The simple gravity return heat pipe (Fig. 10.7) consists of a closed tube containing a volatile liquid in contact with its own vapour at near saturation. When the lower end of the heat pipe

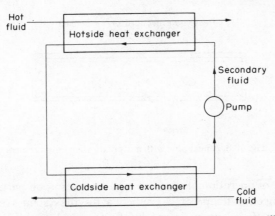

Fig. 10.6. Sketch of the operating arrangement for "run-around" coil.

(the evaporator) is heated whilst the upper end (the condenser) is cooled, the liquid evaporates, diffuses, and convects towards the condenser, and condenses. The combined process of evaporation, vapour movement, and condensation results in a very high rate of heat transfer for very low temperature differences. Capillary wicks may be provided to return the condensate to the evaporator for use in orientations other than the vertical (Fig. 10.8). The capacity of these wicks may be designed to meet the dissipation requirements. Control over the rate of heat transmission may also be achieved by tilting the heat pipe, thus varying the performance of the wick which is, in effect, a capillary pump working against a gravitational head [66]. The geometry of the heat pipe is not limited to a circular "pipe" cross-section. A phase-change panel consists of a flattened heat pipe [67] which may be used, with or without baffles or radiation shields, in the walls of buildings to inhibit heat losses whilst encouraging heat gains. Current usage of these panels is, however, limited to gravity-return systems, and so much development concerning wicking arrangements is required. Alternatively, batteries of heat pipes in parallel, connecting finned surfaces, have been successfully incorporated into air-to-air heat exchange systems.

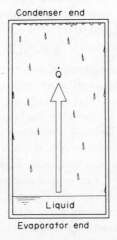

Fig. 10.7. Sketch of a simple gravity return heat pipe.

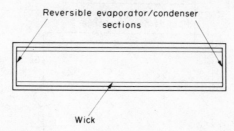

Fig. 10.8. A heat pipe with a capillary wick liquid return.

Because the mode of operation of the heat pipe depends upon partial pressure, and not temperature, differences, no temperature gradient across the system is required theoretically. Thus heat transfer coefficients can approach infinity, enabling thermodynamically down-graded heat to be transferred efficiently. In practice, temperature gradients across the walls of the container are required to allow the conductive heat flows across the system boundaries necessary for continuity of heat flow.

The constructional materials should be selected so as to be compatible with the transfer fluid. Decomposition of the working fluid by direct or indirect action with the wick or the container can form non-condensable gases which impede operation (i.e. the generation of hydrogen by the chemical reaction of stainless steel and water).

A further advantage of the heat-pipe technique is that the collecting surface or evaporator section can be positioned remote from the condenser section. The addition of a pump to aid vapour movement then converts the heat pipe to a simple heat pump.

10.6   HEAT PUMP

Any refrigerator unit is a form of heat pump. The emphasis is, however, nearly always on the cooling capabilities. The heating possibilities inherent in the heat transfer process involved are seldom taken into consideration. Figure 10.9 shows a schematic arrangement for an elementary mechanical vapour compression heat pump system or refrigerator unit. A refrigeration cycle may be considered as a reversed power cycle, heat being received at a low temperature and rejected at a high temperature. If this rejected heat is usefully employed then the cycle may be termed a heat pump.

Mode of operation. Fluid state points 1, 2, 3, and 4 are indicated in the figure. Vapour is compressed isentropically from a low pressure and temperature to a high pressure and temperature from state point 1 to state point 2 where it is condensed to state point 3 by cooling at constant pressure and temperature. The cooling medium (usually water or air) absorbs the latent heat of condensation from the working fluid via a recuperative matrix. The result-ing liquid is then throttled back to the low-pressure side of the system to state point 4. The low-temperature low-pressure liquid then absorbs the necessary latent heat through the walls of a regenerative evaporator to allow it to evaporate back to state point 1. Thus the system forms a closed thermodynamic cycle. The evaporator may absorb heat from refrigerated compartments whilst the condenser rejects heat directly, or indirectly, to the environment, or the condenser may supply heat for a useful purpose (i.e. space heating), whilst the evapor-ator absorbs low-temperature heat from the environment. The former represents a refrigerat-ing application, whilst the latter forms a heat pump.

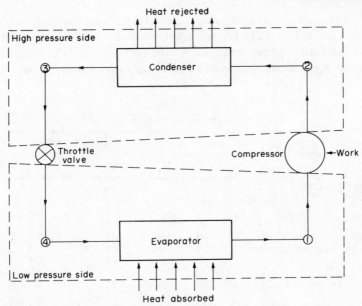

Fig. 10.9. Schematic arrangement of an elementary vapour compression heat pump or refrigeration system.

10.6.1 Coefficient of performance

Because the heat pump, or refrigeration, cycle constitutes a reversed heat engine, the same conditions apply generally for the achievement of maximum performance efficiency. Carnot efficiency for a heat engine is expressed as

$$\eta = \frac{T_1 - T_2}{T_1},\tag{10.21}$$

where T_1 and T_2 are the absolute temperatures at which heat is added to and rejected from the system [68]. When the cycle is reversed and work is done on the fluid to transfer or pump heat between two different temperatures, the efficiency concept cannot be applied, and so an alternative measure, the coefficient of performance (COP) is used. This is defined as the ratio of the heat rejected at the higher temperature to the work input, i.e.

$$\text{COP} = \frac{T_2}{T_1 - T_2}.\tag{10.22}$$

It can be deduced from eqns. (10.21) and (10.22) that, whereas in the heat engine it is desirable to extend the working temperature range to achieve maximum efficiency, for the refrigerator or heat pump the smaller the temperature range the greater will be the COP. Furthermore, because phase changes are involved, the heat should be added to or rejected from the fluid at constant temperatures.

The derivation of expression (10.22) [68] is based upon an ideal reversible process in which, amongst other factors, the working fluid is expanded isentropically between the condenser and the evaporator. The substitution of a throttle valve for the turbine and other inefficiencies in the practical cycle result in an irreversible process with a COP which is less than the ideal Carnot value. Further reductions occur because the evaporation is often continued into the superheat region and the liquid leaving the condenser is subcooled. Fluid friction and compression losses also reduce the COP. Another major limitation is that heat transfers in the evaporator and condenser require temperature differences between the refrigerant and the external heat-carrying medium. Table 10.2 lists comparative values for the coefficients of performance for various systems.

TABLE 10.2. COEFFICIENTS OF PERFORMANCE

Cycle details	Condenser temperature (°C)	Evaporator temperature (°C)	Coefficient of performance
Ideal Carnot	50	10	8.1
	50	4	7.0
	50	−1	6.3
	30	−15	5.7
Freon 12 (no mechanical	50	10	7.2
losses, superheat, or subcooling)	50	4	6.2
	50	−1	5.4
	30	−15	4.8
Practical Freon 12	50	10	5.5
	50	4	4.7
	50	−1	4.1
Ammonia (no losses)	30	−15	4.7
Carbon dioxide (no losses)	30	−15	2.6
Methyl chloride (no losses)	30	−15	4.6
Sulphur dioxide (no losses)	30	−15	4.7
Electrical resistance heating			1.0

10.6.2 **Operational classifications**

Heat pump systems may be classified simply as air-to-air, air-to-water, or water-to-water arrangements, depending upon the heat source (or sink) and the heating or cooling medium.

An air-to-air heat pump system uses outside air as a heat source (or sink when the purpose of the unit is to refrigerate or cool) and air to take up heat from the condenser. Available units are simple and self-contained and there are no problems of water disposal or availability. A high COP is possible when the temperature difference between the outside environmental air and the conditioned internal environmental air is small. Because of this, applications are usually limited to mild climates, as low outside air temperatures reduce the COP and so less heat is available when it is most needed. Another disadvantage of the air-to-air heat pump is the size of the heat exchangers required. Frost accumulation on the outside air heat exchanger matrix can become a problem.

Air-to-water systems, using water to extract heat from the condenser, suffer similar disadvantages, but the sizes of units are reduced by a certain extent.

A water-to-air heat pump uses water as the heat source and air to transfer the heat to the conditioned space. The water can be supplied from a river, lake, well, or other source which has a reasonably constant temperature. The advantage of such a system is the relatively constant COP obtained, which is independent of the outside air temperature, and so application is possible in more extreme climates. Disadvantages include the problems of water supply and disposal.

Systems using water as both heat source and sink have fairly constant coefficients of performance and are much more compact than air-to-air systems.

10.6.3   **The need for advanced heat pump technology**

Despite the inherent thermodynamic advantages of the heat pump, its development in the United Kingdom has been neglected, mainly because of the cheapness and availability of coal and other fuels. The temperate climate experienced tends to make the problem of large-scale heating and cooling of buildings less acute than appertains in colder countries. Un-limited quantities of low-grade heat are, however, available in the earth, rivers, and atmosphere. Using heat pump technology, these sources can be harnessed for an expenditure of comparatively small quantities of high-grade energy. If the direct heating of buildings were replaced by heat pump systems, the savings of valuable fossil fuels would be immense.

Chapter 11

Alternative Energy Sources

11.1 PRIMARY ENERGY SOURCE

The primary source of all energy utilised on the earth, with a few minor exceptions—notably nuclear, geothermal, and tidal power—is radiant solar energy. Fossil fuels represent the accumulation of 400 million years of solar irradiance [4], transformed by photosynthesis into vegetation. Fossil fuels reserves are non-renewable energy resources which will have been completely used up over 2000 years of accelerating exploitation by the middle of the twenty-first century. Energy-planning rationale is influenced by a belief in either one of two possible solutions to the problems of fossil-fuel depletion: (a) that nuclear power from fission or fusion reactors will be able to supply all energy requirements in the near future, or (b) that the only long-term solution lies in the adoption of a solar-based economy.

It may be demonstrated [4] that the crude fission reactors which are currently in operation or at the planning stage, could use up all known uranium reserves in less than 40 years. The more sophisticated fast-breeder reactor in theory produces more fissile material than it consumes. The use of practical fast-breeders could extend the life of raw uranium reserves by about 5000 years. Fast reactors are, however, inherently less safe than the present thermal reactors, suitable sites for such devices are not readily available in the United Kingdom, and the transportation of fuel elements between reactors and processing plants is already causing anxiety. The highly radioactive waste from any nuclear process takes about 800 years to decay to a safe level. The secure storage of waste will thus become a considerable problem. Solutions under consideration include submerging stainless steel cannisters containing vitrified ingots of waste in Antarctic ice, burying them underground, or ejecting them into space. The UKAEA is currently investigating sites for underground storage in the Hebrides, the southern uplands of Scotland, and parts of Cornwall. The 100% safe storage system has not yet been developed, as endorsed by the leak at Windscale in late summer 1976. It is becoming increasingly evident that mankind has not, at the present time, neither the technical expertise nor the facilities to handle any major fast-breeder programme safely. Much research and development into all aspects of nuclear reactor safety, controls, construction and maintenance techniques, and radioactive waste disposal, is required. The fusion reaction system, although inherently safer in many aspects, has not yet been developed. Thus a blind belief in technological salvation is, at least, immature.

The establishment of a solar-based economy in the very near future seems to be the only alternative to a sudden, complete global economic seizure in the short term. It is clear that the rate of fossil-fuel consumption cannot be allowed to continue to increase exponentially.

These reserves are needed for the production of plastics, oil, medicines, and other products as well as for power generation. Neither can the rate of release of energy from any source on the earth be allowed to reach a situation where major changes in climatic conditions, pollution levels, and ecological balances result.

Chapter 4 has shown that $\sim 10^{11}$ MJ of solar energy is intercepted at the surface of the earth every second. Thus a quantity of solar energy which is equal to the entire global annual release of fossil fuel energy falls on the earth every 48 minutes. It follows that if this radiant energy were collected, converted, and utilised with an efficiency of only 0.009%, the world's current rate of fossil-fuel exploitation could be supplied directly from the sun.

Consumption of energy may be reduced by (a) insulating, (b) using heat reclaim devices, (c) developing optimum designs for systems and products, and (d) social reorganisation. If the rate of employment of renewable, non-polluting, solar-based energy was increased to meet the remaining demand, world energy flow would form a closed cycle with the rate of solar influx balancing the rate of energy rejection into space by long-wave radiation. During the brief period of residence on the earth this energy would be utilised to the full. The world would be pollution-free and self-sufficient in energy as long as the sun continues to emit radiation.

11.2 INDIRECT SOLAR ENERGY

The term "solar energy utilisation" usually refers to the direct harnessing and employment of the sun's radiation using thermal collectors. Wind and water power are founded on energy potential differences produced by solar-induced climatic variations.

11.2.1 Wind power

Wind power is mechanical energy which may be used to drive a shaft and so may be converted to electricity without the thermodynamic limitations of heat-based generation. Windmills extract kinetic energy from the wind proportional to the cube of the wind velocity [69]. No machine can convert all the power in an airstream, and the theoretical limit to the efficiency of a rotor is $\sim 60\%$ [70]. Most existing well-designed systems harness 40–45% of the energy available. When the rotor drives an electric generator, the maximum electrical power output varies between 30% and 35%. Large wind-driven machines need very strong supporting structures which must be substantially overdesigned to cope with unusually high wind forces. Because wind power is diverse, and so best suited for small individual installations, an array of smaller devices serving individual systems is often more efficient and reliable. The major objection to the intensive collection of wind power is that the necessary arrays of windmills would be unsightly. Nevertheless, in the solar-based economy, wind machines could be the most efficient electricity producers.

11.2.2 Hydroelectric power

The potential world capacity for hydroelectric power has been estimated as being of the order of 3×10^6 MW [4] of which only about 7% is currently being utilised. Hydroelectric power is often regarded as a renewable resource since inland lakes and reservoirs are filled by the solar-motivated evaporation/rain cycle. In practice the lives of hydroelectric systems are limited by sedimentation and ecological factors. These tend to reduce the long-term cost-effectiveness of many water-based schemes.

11.2.3 Ocean energy

Ocean thermal energy conversion is a solar-based concept in which the constant temperature difference between the warm surface water of tropical oceans and the deeper cold layers returning from the polar regions is used to drive a heat engine [71]. Although conditions are best (i.e. with the largest temperature differences) within $\pm 10°$ latitude of the equator, 80 million square kilometres of relatively calm sea is available within this region. Many possible designs for energy conversion systems exist, but the most promising arrangement uses a closed Rankine cycle with ammonia or propane as the working fluid. The liquid is evaporated at $\sim 21°C$ by being heated by sea water at $\sim 26°C$. After driving a turbine, the fluid is recondensed at $\sim 11°C$ by cold sea water at $\sim 5°C$ drawn from a depth of ~ 800 m. Thus the system operates on a temperature difference of only $\sim 21°C$. Real net efficiencies have been estimated as being of the order of 2.5%, although there is scope for much development, particularly with respect to the heat exchanger equipment.

11.2.4 Biomass energy

All existing reserves of fossil fuel have been prepared by nature via photosynthesis. Until comparatively recently, wood and charcoal fuelled most of man's energy needs. Modern technology is capable of producing quantities of vegetation-based biomass fuels at costs competitive with those of coal products [72]. These fuels can take the form of algae, higher plants, trees, seaweed, or even farm or urban wastes. Whilst it is unlikely that densely populated, industrialised areas have the necessary space for cultivation of non-food-producing crops, possible development sites exist in underdeveloped countries and in regions which are unsuitable for food production. Photosynthesis converts carbon dioxide and water into simple carbohydrates and oxygen using solar radiation absorbed by a plant's chlorophyll. It has been estimated [72] that photosynthetic conversion on the earth harnesses solar energy at a rate which is currently 10 times the total rate of man's use of energy. The energy expenditure in energy-intensive methods of food production are much greater than the energy return produced. Exceptions to this rule occur, however, in the cultivation of some non-food plants such as water hyacinth, Sudan grass, eucalyptus, chlorella, some algae, and bamboo. Yet few of these are being intensively cultivated at this time. Other high-yield crops, such as potatoes or sugar-beet, can be grown and converted into alcohol. The upper limit for the annual production of biomass fuels is about 50 g m^{-2} in fertile tropical regions and about 5 g m^{-2} in Europe. Wood has a calorific value of ~ 15 MJ kg^{-1}, and so a land area of ~ 5000 m^2 would be needed to provide a steady thermal energy output of ~ 7 kW.

The reverse process to photosynthesis is aerobic or anaerobic decomposition [72]. In the former process, bacteria feeding on organic matter in an oxygen-rich environment dissipate heat, carbon dioxide, and water. Anaerobic decomposition occurs in the absence of oxygen and produces combustible methane instead of water. Much of the solar energy stored by photosynthesis is rejected back to the atmosphere during these bacteriological processes. Thus an alternative source of energy lies in the controlled recycling of waste vegetable matter. The material used can take the form of agricultural crop waste, manure, food waste, sewage, paper or wood waste, and either incineration or biodegradation can be employed.

Energy or fuel production processes related to biomass techniques include all light-induced chemical changes such as occur in bleaching, "sun-tanning", or photographic plates. In photodissociation processes, water is dissociated by photolysis, using sunlight, into its components (hydrogen and oxygen) to produce transportable fuels. This is difficult to occasion on a large scale because water, being transparent, does not readily absorb solar energy. There is a research and development need to examine changes in the chemical and radiative properties of water which take place with the addition of colourants.

11.3 TIDAL AND WAVE ENERGY

The tides depend upon the kinetic and potential energy of the earth–moon–sun system. It has been shown that, whilst the total energy available in the rise and fall of the tides is about 3×10^6 MW, practical sites existing on world coastlines could extract only $\sim 64\,000$ MW [4]. In the United Kingdom, the most suitable region for the erection of a tidal barrier is at the Severn estuary where a possible 300 MW could be produced. The capital investment would, however, be considerable and such a scheme may inconvenience shipping. The life of a tidal electricity generating system is limited by silting. Because the power output would be variable, installations should incorporate some means of energy storage, either as pumped water or as electricity.

Wave energy utilisation involves harnessing the random fluctuations of surface waves. Many crude oscillatory devices have been proposed to achieve this, but there are a great number of technical problems to be overcome before such systems become economically feasible. If the power in the waves around the entire British coastline were collected, $\sim 22\%$ of the total UK energy requirements would be met [12].

11.4 GEOTHERMAL ENERGY

The earth has a mass of 5.976×10^{24} kg. Assuming a mean specific heat of $1000\,\mathrm{J\,kg^{-1}\,K^{-1}}$, it is estimated that a $0.1°C$ general reduction in its mean temperature would release 5.976×10^{24} MJ of thermal energy. This amount could supply the world's current needs $(0.287 \times 10^{-15}$ MJ per annum) for 2 million years.

Energy from existing hot springs in volcanic regions could, if entirely tapped, provide 120×10^6 MW years [4] of energy†: 1000 MW is already being obtained in this way. Geothermal fields decay in time and so cannot be regarded as renewable resources. Deep drilling to molten rock could produce further sources into which water might be injected to produce steam. Sadly, existing technology is insufficiently sophisticated to achieve this, and so future prospects of tapping energy from the earth's core are inestimable.

11.5 DIRECT UTILISATION OF SOLAR ENERGY

Direct solar energy is a clean, cheap, non-polluting source which is available to all. Even at high latitudes, longer days partially compensate for lower concentrations of solar radiation. Solar radiative flux is extremely dilute ($1.35\,\mathrm{kW\,m^{-2}}$ before atmospheric absorption, reflection, and scattering), having an approximate average annual intensity (in the absence of cloud cover) in the United Kingdom of $120\,\mathrm{W\,m^{-2}}$ (Chapter 6). Thus it is difficult to achieve economies of scale in solar energy collection technology. The weak, low-grade,

---

†Present indications are that geothermal energy from hot springs could produce only $\sim 1\%$ of global energy requirements for about 50–100 years [4].

energy may, however, be effectively used in small units to accomplish individual tasks which require relatively small amounts of low-grade heat [44] such as providing hot water, house-heating, distillation, drying, or refrigerating (using an absorption cycle).

11.5.1 Types of solar energy collectors

Giant satellites have been proposed to intercept the higher power radiation at the fringe of the atmosphere and to convert this to high intensity microwave radiation to be beamed to the earth's surface, where conversion to electrical energy could take place. Unfortunately, this technique is far beyond present capabilities, and it is probable that the net energy which would have to be expended to establish such a system could outweigh much of the energy collected during the system's lifetime.

Ground-based direct solar-radiation-concentrating collectors achieve an increase in radiative flux and hence surface equilibrium temperatures via magnification using the ratios of areas. Devices may be cylindrical, parabolic, or paraboloidal. Simpler concentrating arrangements may be constructed using individual plane mirrors. The ratio of solar inter-ception area to heat loss area, which is unity for a flat-plate collector, may be increased considerably in concentrating systems. The radiation may be focused on to a boiler surface, a thermionic device, or a furnace. Major disadvantages are that the large mirrors required must follow the sun, and wind forces often render the cost of "sun-seeking" mounts prohibi-tive. Large areas of land are often required for large installations, and most designs are expensive to build.

11.5.2 Flat-plate collectors

Whereas focusing collectors can only harness the specular component of the sun's rays, the flat-plate collector also collects diffuse sky radiation which can be up to 30% of the total irradiance. Furthermore, because these also operate at lower temperatures than do concen-trators, heat losses are reduced. Although concentrators have higher overall thermodynamic efficiencies, flat-plate collectors offer the most possibilities for widespread application because of their cheapness and convenience.

As the temperature of the plate rises, its conversion efficiency, from incoming sunlight to sensible heat addition to a secondary stream of air or liquid, decreases because of increased heat losses from the surface of the plate [44]. The temperature of the plate, and hence the outlet fluid temperature, can be raised without reducing the conversion efficiency by:

   (i) optimising the number of air-spaced glass cover-plates with respect to surface air flows and the radiant characteristics of the glass;
  (ii) optimising the distance between cover plates;
 (iii) evacuating the gaps between cover plates;
 (iv) treating the collector plates with non-reflecting coatings (such as heavy metal soap);
  (v) reducing convective losses;
 (vi) improving the back insulation of the collector;
(vii) making the black coating of the collector wavelength selective.

Selective absorbers absorb short-wave radiation well but emit longer waves poorly. They usually consist of polished metal surfaces coated with thin deposits of black salts such as the oxides of nickel or copper. These films have high absorptivities (~ 0.9) for solar radiation. If the layer is thin, however, it is found to be transparent to radiation longer than the coating thickness. Then, for the long-wave radiation appropriate to absorber temperatures, the emissivity is close to that of the metal surface beneath (i.e. about 0.1 for aluminium). The ratio of a surface's absorptivity to solar radiation to its emissivity for long-wave radiation $\mathscr{A}_s/\varepsilon_{temp}$ is thus a useful parameter for classifying solar-collecting surfaces (Table 11.1), although the convective resistance at the surface plays a dominant role. Whilst some of the values given in Table 11.1 appear to be attractive, the absorber surface should also have a high individual value of absorptivity and the material used should have a high thermal capacity

TABLE 11.1. TYPICAL VALUES OF THE RATIO OF ABSORPTIVITY TO SOLAR RADIATION TO EMISSIVITY AT 25°C FOR VARIOUS SURFACES

Material	$\mathscr{A}_s/s_{temp}$	Material	$\mathscr{A}_s/e_{temp}$
Polished zinc	23.00	Brick	0.65
Polished silver	11.00	Black paint	1.00
Polished aluminium	7.50	White paper	0.25
Polished iron	7.50	White paint	0.15
Asphalt	1.00	Glass	0.01

so that more heat can be absorbed for a given temperature rise. Surface-to-air convection losses may be reduced by using cover plates or by constructing cellular pockets of still air at the surface. In order to intercept the maximum total solar intensity, flat-plate collectors, which are not "solar-seeking", should be tilted towards the equator at an angle which is 15° greater than the latitude of the location in summer and should be vertical in winter. Some form of thermal energy storage is a desirable feature of solar energy collection systems to ballast seasonal and diurnal variations in radiant intensity. The capacities of these stores should be of the order of 0.1 m^3 (water) per m^2 of collector surface [44]. Successful arrangements for solar energy collection and storage will be compact, simple, reliable, and cheap.

Equilibrium temperatures for flat-plate collectors are readily calculated using basic heat transfer analyses providing all the variables are known accurately. For an irradiance of 800 W m^{-2}, in the absence of wind, an uncovered blackbody absorber would reach $\sim 70°C$, a glass cover-plate would increase this to $\sim 113°C$, an uncovered selective absorber could reach 160°C, and a selective surface covered by a glass plate could attain an equilibrium temperature of $\sim 190°C$ [44]. These values apply when no heat is extracted. A small amount of heat at a high temperature, or a large amount of heat at a lower temperature, can be drawn off. It has been calculated [44] that the efficiency of a flat-plate collector decreases from $\sim 60\%$ for an exit fluid temperature of 40°C to $\sim 40\%$ if the fluid leaves at 60°C.

11.5.3 Conversion to electricity

Photovoltaic, or solar, cells operate on photoelectric effects by which light, falling on a specially prepared boundary of certain pairs of substances (e.g. copper or cuprous oxide), produces a potential difference across the interface. The most efficient devices available use silicon or germanium semiconductors. The power output from silicon cell systems can produce ~ 2 MJ m^{-2} (at $\sim 11\%$ efficiency). Existing components are prohibitively expensive, and so much research and development towards cheaper manufacturing techniques is required.

Thermoelectric devices operate on the Seebeck effect: if a closed circuit is formed between two dissimilar metals an electrical current flows in the circuit when the two junctions are maintained at different temperatures. This is the principle of the thermocouple temperature-measuring device. The open circuit voltage developed, Δv, is proportional to the temperature difference ΔT, i.e.

$$\Delta v = b_{12}\Delta T \tag{11.1}$$

where b_{12} is the relative Seebeck coefficient for metals 1 and 2. Typically, $b \simeq 0.00005 \text{ V K}^{-1}$ for metals and 0.00022 for semiconductors. Thus efficiencies are very low at present, and further progress in the design of thermoelectric generators awaits the development of materials with higher Seebeck coefficients.

11.5.4   **The solar dome**

It has long been realised that if a building were insulated against heat loss and completely surrounded by glass panels, the solar pick-up would be such as to maintain the enclosed system intolerably hot for most of the year, even in UK latitudes. Energy would then be required in quantity to cool the interior. It would, however, be possible to maintain comfort conditions using optimally designed structures with proper air mixing and distribution systems together with a thermal storage arrangement. External temperature and solar irradiance fluctuations may be smoothed out using thermal rectifiers and wall storage.

11.5.5   **Solar cooling**

A refrigeration cycle may be considered as a reversed power cycle, heat being received at a low temperature and rejected at a high temperature. If the heat rejected is usefully employed, the cycle is termed a heat pump.

11.5.5.1   VAPOUR COMPRESSION REFRIGERATION SYSTEMS

Figure 10.9 showed a schematic arrangement for a vapour compression system operating on a reversed Carnot cycle. The working fluid, the refrigerant, passes through four state points in a closed cycle. Between state points 1 and 2 the fluid is compressed isentropically (i.e. with theoretically no transfer of heat across the system boundary and therefore under constant entropy) from the low pressure and temperature which exists in the evaporator to the higher pressure and temperature which prevails in the condenser. The vapour is then condensed to state point 3, giving out latent heat. The resulting liquid is re-expanded isentropically to the evaporator conditions at state point 4. The low-pressure liquid then evaporates, absorbing the necessary latent heat from the surroundings. The cycle requires a net input of mechanical energy for motivation. Practical vapour compression refrigeration cycles replace the isentropic expander (i.e. a turbine) by a throttle valve which achieves the drop in pressure and temperature at constant enthalpy. Suitable refrigerants include ammonia, carbon dioxide, methyl chloride, sulphur dioxide, and the various Freons.

11.5.5.2   ABSORPTION SYSTEMS

The work of compression can be reduced significantly if the vapour is dissolved in a suitable liquid before being compressed. Then the vapour can be drawn off at the higher pressure and throttled and evaporated in the usual way. Thus the simple absorption refrigeration system substitutes an absorber, a generator, a liquid pump, and a throttled liquid feedback system for the vapour compressor [74]. Figure 11.1 shows the schematic circuit for a simple lithium-bromide–water refrigerator. This system predominates in air-conditioning practice because water evaporating at $\sim 4°C$ is the refrigerant, and so hazards associated with leaks are

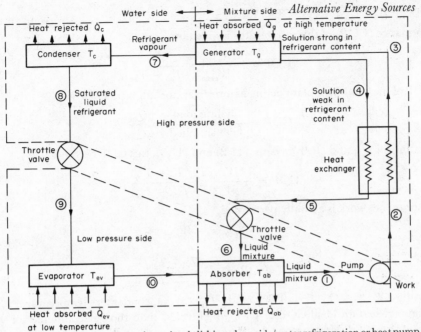

Fig. 11.1. Schematic circuit representing a simple lithium-bromide/water refrigeration or heat pump system.

reduced. The absorbent lithium bromide is a solid which forms liquid solutions when it is mixed with sufficient water, the solubility of water in lithium bromide being higher the lower the temperature of the mixture. Between state points 1 and 2 (Fig. 11.1), vapour is passed into the absorber and dissolved in a water–lithium–bromide mixture, forming a solution which is "strong" in water content. Heat is rejected from the absorber to the surroundings as the vapour condenses into the mixture. Cooling arrangements can be designed to serve the condenser and the absorber simultaneously, and so those two components are often maintained at the same temperature. The liquid solution is then pumped up to the generator and condenser pressure and heated in two stages: first in a heat exchanger which recoups some of the heat in the weak solution returning from the generator, and then in the generator itself. This causes the refrigerant to boil off from the mixture absorbing the latent heat of evaporation required from the environment at high temperature. Thus relatively high grade heat (at ~ 100°C) must be supplied to the generator to cause this to occur. The pure vapour then passes to the condenser. The overall work input is small but the cycle is predominantly motivated by the supply of heat to the generator.

An ideal coefficient of performance for an absorption refrigeration system may be estimated by considering the system to comprise of two thermodynamically reversible machines which, taken together, perform the function of the absorption plant: (a) a reversible heat engine which receives a quantity of heat Q_g at T_g (Fig. 11.1) (the generator temperature) and rejects heat Q_{ab} at T_{eb} (the absorber temperature) whilst producing a quantity of work W_{gab}: the Carnot efficiency of this heat engine is given by

$$\eta = \frac{W_{gab}}{\dot{Q}_g} = \frac{T_g - T_{ab}}{T_g}; \tag{11.2}$$

and (b) a reversible refrigerator which receives a quantity of heat $\dot{Q}_{ev}$ at T_{ev} (the evaporator temperature) and rejects heat at T_{ab} whilst absorbing a quantity of work W_{evab}. The coefficient of performance (COP) of this refrigerator is given by

$$\text{COP} = -\frac{Q_{ev}}{W_{evab}} = \frac{T_{ev}}{T_{ab} - T_{ev}}. \tag{11.3}$$

The COP of the combined plant acting as a refrigerator can be defined as

$$\text{COP} = \frac{\text{``cooling load''}}{\text{heat supplied}} = \frac{Q_{ev}}{Q_g}. \tag{11.4}$$

Substituting for Q_{ev} and Q_g from eqns. (11.2) and (11.3), this COP becomes

$$\text{COP} = -\frac{W_{evab}T_{ev}(T_g - T_{ab})}{W_{gab}T_g(T_{ab} - T_{ev})}, \tag{11.5}$$

and, since no net work is actually performed,

$$W_{gab} = -W_{evab}$$

and so

$$\text{COP} = \frac{T_{ev}(T_g - T_{ab})}{T_g(T_{ab} - T_{ev})}. \tag{11.6}$$

Typically, $T_{ev} = 4°C$, $T_g = 95°C$, and T_{ab} ($=$ the condenser temperature) $= 40°C$. These figures yield an ideal COP of 1.15, which is less than those encountered for vapour compression systems. Most practical systems operate with coefficients of performance less than unity, and so a cheap source of heat is required. The solar refrigerator uses solar irradiation incident on the surface of a flat-plate collector to supply the necessary heat at the moderately high temperatures required. Arrangements driven by waste heat have also been developed. Because the vapour generation must be accomplished at constant temperature for most efficient operation, it is advisable to rectify and ballast incident energy fluctuations using latent or sensible heat stores.

11.6   THERMAL RECTIFICATION

Thermal energy from low-grade intermittent sources will in future be increasingly harnessed for higher grade usage. The conversion processes involved often require collection systems incorporating thermally rectifying input channels to an energy storage facility. A thermal rectifier is a device [67] which has a high resistance to heat flow in one direction across it, together with a low resistance for the reverse direction of heat flow. Such devices are necessary for maintaining constant temperatures in systems which are intermittently exposed to or isolated from a source (or sink) for heat and may be employed in energy collectors or ballast arrangements. Simple rectifiers include the heat pipe, solar ponds, thermally distorting joints, inclined cavities, as well as those which are wavelength-selective in transmission or absorption. Practical designs may incorporate two or more types of rectifiers in series or parallel to produce high or low overall resistance assemblies or to extend the range of conditions over which the system is required to function.

Intrinsic to any thermal energy collection system there are components operating as an interface with an energy source, a thermal rectifier, and a heat store. The energy flow chart (Fig. 11.2) illustrates interconnections between these basic elements for various energy sources.

Whereas mechanical (i.e. non-return valves) and electrical (i.e. diodes and transistors) rectifiers are commonplace, the thermal counterpart has received little direct attention, being often assumed to be part of the collector device or the storage facility. The intermittent behaviour of many sources (i.e. solar, waste, and excess heat) renders successful rectification necessary. The ideal rectifier would behave as a superinsulator or a superconductor depending upon the direction and mode of thermal energy passing through it. The heat captured by the device must be transported to an adequate temperature-controlled heat store either directly or via a secondary transfer fluid.

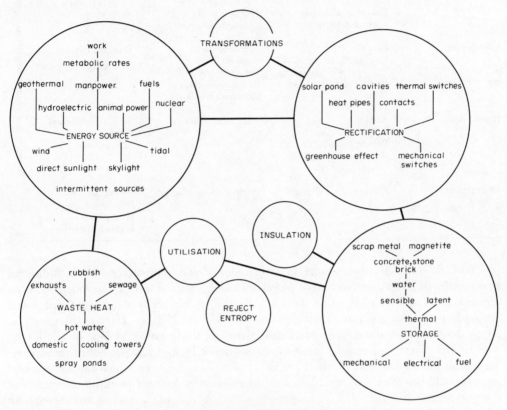

Fig. 11.2. Energy flow chart.

11.6.1 **Inhibition and enhancement of heat transfer**

The relevant transmission modes of low-grade heat are by conduction in solids and stagnant fluids, by convection in moving fluids, by radiation between surfaces, and by mass transfer combined with changes of phase. Table 11.2 summarises methods which may be used to obtain increased or reduced heat transport by each mode.

Heat transfer by solid conduction may be effectively impeded utilising contact resistances [58]. Superinsulations can consist of stacks of hard, thin, solid laminations with rough, wavy surfaces; an increase in the load applied normal to the planes of the interfaces reduces the overall thermal resistance. Good thermal contact between solids is achieved by the addition of high conductivity oils or greases, or soft interfacial shims with smooth flat surfaces.

Stagnant gases at atmospheric pressure have low thermal conductivities. Most thermal insulants trap air in interstices rendering the insulating effectiveness proportional to the amount of stagnant air contained therein [41]. Liquids have higher conductivities, and so wet insulations have better heat transport properties. As the Grashof number increases

TABLE 11.2. INHIBITION AND ENHANCEMENT OF HEAT TRANSFER AND THERMAL RECTIFICATION

Mode of heat transfer	Inhibition	Enhancement	Type of rectifier	
Solid conduction	Eliminate solid bridges Reduce areas for heat flow Increase lengths of thermal paths Decrease solid conductivities	Provide continuous paths Increase areas for heat flow Decrease distance between source and sink Increase solid conductivities	Contacts subjected to thermal distortions or varying applied normal loads	
Convection	Low Gr Evacuate Hold gases in pockets Reduce areas for heat flow	Speed up flow High Re Pressurise Reduce viscosities Promote turbulence	Damped cavities	Vasometer regulation and sweating
Radiation	Shields Low emissivities Prevent "seeing" Reduce areas for heat flow	High emissivities	Selective absorbers Filters "Greenhouse effect"	
Mass transfer	Prevent phase changes	Promote phase changes	Heat pipe Solar pond Heat pump Diffusion barriers	

($\gtrsim 2000$), fluid within a system starts to move under buoyancy forces resulting in additional heat transfer by natural convection. A pump or blower promotes forced convection and even higher rates of heat transfer, whilst the onset of turbulence in the flow (at high Grashof and Reynolds numbers) generates very high heat transfer coefficients. Heat exchangers (i.e. recuperators, regenerators, run-around coils) are designed to maximise heat dissipation by convection by increasing transfer areas, increasing flow velocities, and promoting turbulence. To inhibit convective heat transfer, the exposed surfaces between which fluid movement occurs should be reduced in area, flow velocities should be lowered to minimum values, or the system may be evacuated. A reduction in the mean flow channel radius through an insulant results in lower Grashof numbers. Thus powders, foams, and fibres often form the basis of thermal insulations. The optimum cavity separation for highest thermal resistance is of the order of 19 mm [62]. The height of an insulating cavity should be small to avoid turbulence, but care must be taken not to produce additional solid transmission bridges by the addition of baffles. Stack effects may be utilised to promote convection with dampers to provide some control.

Transfer of heat by thermal radiation needs no intervening substance between two surfaces but may be effectively minimised by reducing surface areas by introducing shields and bends to prevent temperature potentials "seeing" one another or by using low emissivity surfaces. Thermally "black" surfaces produce the highest rates of radiant heat exchange. The rougher a surface, the closer it approaches the blackbody because radiation is absorbed by inter-asperity multiple reflections and absorption [22].

Although vapour diffusion in itself adds little to heat transfer rates, very high overall rates across small temperature gradients can be accomplished when phase changes occur [23] (i.e. evaporation and condensation). Any insulated system must be designed so that phase changes are prevented at its boundaries. Vapour barriers may be used to contain condensable vapours within the system at temperatures above their dew-points.

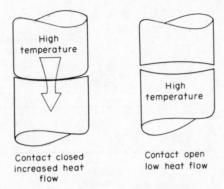

Fig. 11.3. Thermal rectification caused by distortions at solid contacts.

11.6.2   **Existing devices**

A thermally distorting contact between two dissimilar solid materials has different resistances to solid conduction according to the direction of heat flow across it (Fig. 11.3). Cavities containing air can be caused to offer different rates of convective heat transfer for each heat flow direction by the addition of baffles or dampers (Fig. 11.4). In the glass-covered flat-plate solar energy collector a large proportion of the incident solar energy which passes through the glass is absorbed by the collector plate which reradiates photons of predominantly longer wavelengths to which the glass is more opaque (Fig. 11.5). A selective coating is also a form of thermal rectifier, having a low resistance to radiant absorption but a high resistance to emission. A simple gravity-return heat pipe has an appreciably smaller thermal resistance in the direction of the vapour stream, when evaporation and condensation aid the heat transfer process, than in the reverse direction. The capillary-wick system is basically reversible, but techniques may be employed to inhibit normal heat pipe operation in one direction. These include deliberately choking the evaporator or condenser sections with non-condensable gases (Fig. 11.6), draining the condensate from the wick, or arranging that the working fluid freezes at the condenser side [67]. Heat pipes designed for unidirectional applications have been described as "thermal diodes" since they transmit heat very effectively in one direction but virtually not at all in the other.

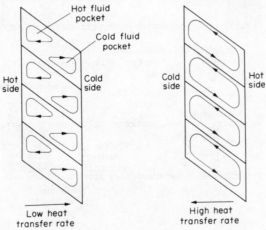

Fig. 11.4. A thermally rectifying air cavity.

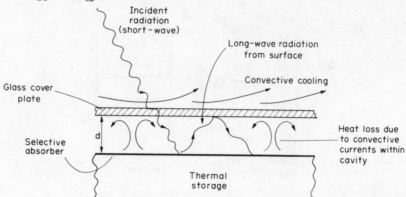

Fig. 11.5. A glass-covered flat-plate solar energy collector.

Heat-rectifying roofs have been described [75] which contain eutectic salts in a cardboard honeycomb. These have been used successfully in the Sahara Desert to exclude solar radiation and to absorb excess room heat during the daytime. During the day, the salts melt in the ceiling as warm air rises in the room. At night the molten salts are saturated with a working fluid which absorbs latent heat and boils off to condense on the cool inner surface of the roof. As a result, the salts freeze again, ready to absorb heat from the room the following day. A reverse procedure can be used to warm a room in winter.

Alternative rectifying systems can easily be envisaged, and some thermal switches and mechanical designs have been described [67]. Invariably, greater efficiency is attained at the cost of additional complexity. Although it is difficult to simultaneously inhibit direct heat flow whilst encouraging reversed heat flow by the same mode of heat transfer, successes in thermal rectifier development can be expected where one mode is enhanced whilst others are discouraged. The cellular-surface solar energy collector, or radiative cooler, where radiative

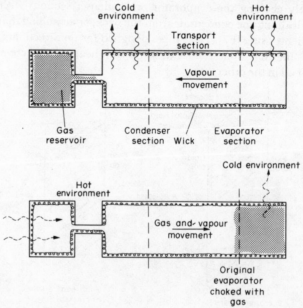

Fig. 11.6. A heat pipe thermal diode using gas to choke the evaporator section for one direction of heat flow.

exchanges are unimpeded but convective movements are minimised by the layer of still air held at the surface, illustrates this principle.

11.7 **THE SOLAR POND**

This consists of a pond of water fitted with a cover plate if desired. Solar or other energy is received by an absorbing tray at the bottom of the pond. If suitable salts are added to the water, the warmer fluid can be caused to fall to the bottom of the pond, the solubility of the salts in the water being greater at higher temperatures producing a denser mixture.

With the hotter, stronger solution at the bottom, the salt component diffuses to the top surface where the concentration is weaker. It thus becomes necessary to periodically remove from the top of the pond and to re-introduce it at the bottom. The so-called "solar pond" represents the cheapest combined collector and storage system for large-scale applications, possessing a large inherent storage capacity because of the high specific heat of water. It is, however, a low-temperature system (although temperatures as high as 100°C have been recorded) and so is limited thermodynamically to possess a low overall efficiency.

11.8 **ENERGY STORAGE**

Because of the intermittent nature of alternative energy sources (i.e. waste heat, direct solar energy, and wind, wave, and tidal power) and the fluctuating requirements of systems to which this energy is to be supplied, it is necessary to include facilities for energy storage and ballast in collection systems. The efficiency of an energy store depends upon the grade and quality of energy which it is capable of accumulating, the possible duration of storage, and the ease by which energy may be added to or subtracted from the system. Primary factors influencing design and selection include capital and runnning costs in both energy and monetary terms. Energy accumulators may be classified as follows:

(a) mechanical storage systems which include flywheel kinetic energy storage, pumped liquid storage, and compressed gas storage arrangements;

(b) chemical storage systems which include fossil and biomass fuels and combustible liquids;

(c) electrochemical storage which includes batteries and accumulators;

(d) thermal storage systems by which heat energy is stored in the sensible or latent heat of single substances, mixtures, or compounds [75].

When the alternative energy source is manifested as mechanical power (wind, wave, or tidal power) the kinetic energy available may be stored in part by using either a rotational system or by pumping. Mechanical energy storage up to 360 MJ is currently possible with practical flywheel designs. Carbon fibre reinforced plastics are being investigated to produce high-strength, high-inertia systems with mass concentrated at a large radius. Small-scale kinetic energy flywheels are being employed in various fields, including high-performance hoists and haulage systems, heat engines, and for braking vehicles. The explosive break-up of large-scale flywheels is an inhibiting factor. The total energy charge at 100 MW for 12 h is equal to about 100 tons of high explosive [76]. Mechanical energy may also be stored in underground caverns by pumping on compressed air [77]. This mode of storage is most suitable for electrical generating stations to ballast for off-peak loads. Alternatively, water can be pumped to a higher level reservoir to drive water-turbines when reclaim is required. All reservoirs suffer mass, and hence potential energy, loss by evaporation.

The chemical storage of fossil and biomass fuels has already been discussed. An alternative energy-to-fuel system produces hydrogen by the electrolysis of water or by the decomposition of metallic hydrides. Major drawbacks to this type of energy storage are the low storage temperatures (~ 21 K) which must be maintained by an external energy supply and that, because the density of liquid hydrogen is low (specific gravity $= 0.07$), large storage volumes are required.

Existing electrochemical storage devices are efficient but expensive, bulky, and have relatively short working lives. Much development is required, therefore, before electrochemistry can be applied for large-scale energy accumulation.

11.8.1 **Thermal accumulators**

Storage of heat is usually limited to low-grade short-term thermal energy in transient behaviour. Because the thermal energy available is dilute and variable, collector arrangements must either cover large areas or the associated thermal store must operate on a trickle charge. The 40% of the total fuel consumed annually in the United Kingdom for space heating could be appreciably reduced if thermal storage were to be recognised as being economically justifiable and hence more widely employed [78]. A radiative cooler takes the form of a solar collector operating in reverse (i.e. by long-wave emission to the black vault of the sky at night). Thus a "cold" store may also be produced which may be used for the thermal sink of an air-conditioning system or for a heat engine operating on low-grade energy.

The widespread use of small, properly designed thermal stores could eliminate the need to expend fuel for heating buildings at night in the United Kingdom during the months from May to October [78]. Long-term (e.g. semi-annual) storage is also possible but requires much higher degrees of thermal insulation and larger volumes for the storage medium. Sensible heat stores accumulate sensible heat by raising the temperature of a material with a large thermal capacity, whilst latent heat stores employ phase changes to store latent heat at constant temperature.

11.8.1.1 SENSIBLE HEAT STORES

The simplest form of heat store consists of an inert body whose temperature rises when heat is absorbed and falls when heat is withdrawn. The transient characteristics depend upon its time constant. The higher the thermal diffusivity the more rapid the thermal response. The response of materials with low thermal diffusivities (i.e. building materials) may be accelerated using greater heat transfer areas or by reducing boundary thermal resistances. Water tanks, bricks, concrete, stone, and magnetite all require quite large masses to achieve useful thermal accumulation. Water is the cheapest and most convenient of these, being transportable and having a high thermal capacity. Its use is, however, restricted by its low boiling-point.

Time constants. Assuming that, at any instant, the temperature is uniform throughout a sensible heat store, its time constant Θ_{sb} may be defined as

$$\Theta_{sb} = \frac{\rho c_p V}{UA}, \tag{11.7}$$

where c_p is the specific heat of the storage material ($J\,kg^{-1}\,K^{-1}$), ρ its density ($kg\,m^{-3}$), V the volume of the store (m^3), U the effective overall boundary heat transfer coefficient ($W\,m^{-2}\,K^{-1}$), and A the surface area of the boundary (m^2). The time constant represents the period for the temperature excess of the storage medium over that of its environment to decrease to 36.8% of its initial value (cf. Chapter 2). Thus the higher the boundary resistance the greater will be the duration for which heat is stored.

A sphere has the least surface area for a given volume: for this

$$\frac{V}{A} = \frac{r}{3} = 0.205 V^{1/3}, \tag{11.8}$$

but this shape is relatively expensive to construct. A cylinder of diameter $2r$ and optimum length $2r$ has

$$\frac{V}{A} = \frac{r}{3} = 0.178 V^{1/3}, \tag{11.9}$$

whereas for a cube of side L (representing the shape most easily manufactured)

$$\frac{V}{A} = \frac{L}{6} = 0.166 V^{1/3}. \tag{11.10}$$

Spatial constraints, however, often require the store to be rectangular. Then for such a system with sides, L_1, L_2, and L_3,

$$\frac{V}{A} = \frac{V}{2(L_1 L_2 + L_2 L_3 + L_3 L_1)}$$
$$= \frac{1}{2}\left(\frac{1}{L_1} + \frac{1}{L_2} + \frac{1}{L_3}\right)^{-1}. \tag{11.11}$$

Having chosen the shape for the storage system, there remains the problem of selecting a cheap, convenient, and safe material to act as the storage medium. Water is usually the cheapest and most convenient substance available for sensible heat storage (Table 11.3) although oils have been used for higher temperature systems. The temperature of the store must always be higher than the least temperature at which the heating fluid needs to be introduced into the space to be heated, but it must also be lower than the temperature of the energiser, e.g. a solar-energy or waste-heat collector.

The Boundary Resistance. If water in its liquid phase is used as a storage medium, a structural housing strong enough to contain it does not normally contribute appreciably to the boundary thermal resistance. Additional insulation must therefore be applied (Table 11.4). The

TABLE 11.3. THERMAL CAPACITIES AT $20\,^{\circ}$C OF SOME TYPICAL SPECIMENS OF COMMONLY AVAILABLE MATERIALS

Material	Density ρ (kg m^{-3})	Specific heat c_p (J kg^{-1} K^{-1})	Volumetric thermal capacity ρc_p (10^6 J m^{-3} K^{-1})
Clay	1458	879	1.28
Common brick	1800	837	1.51
Sandstone	2200	712	1.57
Wood	700	2390	1.67
Concrete	2000	880	1.76
Glass	2710	837	2.27
Aluminium	2710	896	2.43
Iron	7900	452	3.57
Steel	7840	465	3.68
Gravelly earth	2050	1840	3.77
Magnetite	5177	752	3.89
Water	988	4182	4.17

TABLE 11.4. PROPERTIES OF TYPICAL DRY SPECIMENS OF SOME COMMON INSULANTS AT $20\,^{\circ}\text{C}$

Material	Effective thermal conductivity k (W m^{-1} K^{-1})	Thermal resistance of 10 mm thick slab R (m^2 K W^{-1})
Concrete	0.510	0.20
Brick	0.480	0.21
Gravel	0.300	0.33
Aerated concrete	0.150	0.66
Fibreboard	0.060	1.67
Paper	0.060	1.67
Wool blanket	0.043	2.32
Glass wool	0.040	2.50
Cellulose wadding	0.038	2.63
Fibreglass	0.037	2.70
Mineral wool	0.036	2.78
Expanded polystyrene	0.036	2.80
Kapok	0.035	2.86
Stagnant air	0.026	3.84

thermal resistances due to the external air boundary layer and the internal fluid layer are also negligible compared with that of a thick layer of insulant. Thus the boundary resistance may be approximated by

$$R = \frac{\delta}{k} = \frac{1}{U}. \tag{11.12}$$

The thermal resistance of the majority of insulants is dictated primarily by their ability to hold air stagnant within themselves by having sufficiently small pockets (< 2 mm dimension for near ambient temperatures). According to how well they achieve this, their insulating effectiveness tends towards that of still air ($k = 2.6 \times 10^{-2}$ W m^{-1} K^{-1} at 20°C). The selection of a suitable material, however, also depends upon price, availability, strength, etc. A thermal conductivity of 3.5×10^{-2} W m^{-1} K^{-1} is assumed for the insulant adopted in the present calculations. It is also assumed that this insulant is uniformly applied over the surface of the thermal store.

In practice the non-uniformity of boundary insulation and the introduction of thermal bridges arising from inlet and exhaust piping (as well as other phenomena such as stratification in the water), results in an overall thermal resistance considerably less than that which would be predicted from eqn. (11.12). Levels of insulation and isolation should be increased appropriately to compensate for these additional partial thermal short circuits once the initial design parameters have been estimated from an analysis such as is presented here.

By making the appropriate substitutions in eqn. (11.7), time constant in seconds for a cube of side L becomes

$$\Theta_{sb} = 1.98 \times 10^7 L\delta. \tag{11.13}$$

Heat stored. The available sensible heat stored is given by

$$Q = \rho C_p V \Delta T \tag{11.14}$$

which for a cube of side L becomes

$$Q = 4.17 \times 10^6 L^3 \Delta T. \tag{11.15}$$

The total amount of heat dissipated from the store during the interval $\Delta t = \Theta_{sb}$ (i.e. corresponding to 63.2% of the initial available heat stored) is

$$\dot{Q}_{\Theta_{sb}} = 2.64 \times 10^6 L^3 \Delta T, \tag{11.16}$$

where ΔT is the initial temperature excess (i.e. that at $t = 0$).

Effect of accumulator capacity on insulation levels. Equation (11.13) indicates that the thickness of insulant required for a given time constant is inversely proportional to the size of the store, i.e.

$$\delta = \left(\frac{\Theta_{sb}}{1.98 \times 10^7 L} \right).$$

For a given initial temperature excess, the total amount of heat dissipated during the period $t = 0$ to $t = \Theta_{sb}$, i.e. $\dot{Q}_{\Theta_{sb}}$ is directly proportional to the third power of the dimension of a side of the cube (eqn. (11.16)).

Thus two cubical heat stores with sides L_1 and L_2 will have the same time constant for the same initial temperature difference ΔT if

$$\frac{\delta_1}{\delta_2} = \frac{L_2}{L_1} = \left(\frac{\dot{Q}_{\Theta_{sb2}}}{\dot{Q}_{\Theta_{sb1}}} \right)^{1/3}. \tag{11.17}$$

Equation (11.17) illustrates the advantages of scale on the amount of heat which can be dissipated and the appropriate insulation thickness, e.g.

if $\qquad L_2/L_1 = 10,$ then $\qquad \dfrac{\dot{Q}_{\Theta_{sb2}}}{\dot{Q}_{\Theta_{sb1}}} = 1000$ and $\delta_2/\delta_1 = 0.1,$

i.e. a tenfold enlargement in linear dimensions increases the mean dissipation one-thousand fold whilst only one tenth of the insulation thickness is required. Because the ratio of surface area A to capacity V decreases as the accumulator size increases, only relatively modest insulant thicknesses are necessary for large systems. This explains why domestic hot-water tanks and similar small-scale thermal stores need relatively thick layers of insulant to achieve even short time constants.

Sensible heat storage system behaviour. Equations (11.13) and (11.16) and Fig. 11.7 derived from these equations for an initial temperature excess ΔT of 50°C describe the effects of sensible heat accumulator dimensions and insulant thickness on the amount of heat stored and the time taken for a 63.2% reduction in store temperature.

Example 1. Diurnal–noctural environmental temperature fluctuations. In order to allow an average 5 kW rate of dissipation throughout the night, 216 MJ of thermal energy must be released during 12 h. This should represent 63.2% of the total heat stored initially at $\Delta T = 50°C$, i.e 342 MJ. The design chart (Fig. 11.8) indicates that a cube of water of side 1.16 m with an external layer of insulant only 1 mm thick is required to satisfy this condition.

Example 2. Long-term thermal storage. As a rough approximation an average domestic building might require 5 kW continuously for six winter months to offset transmission losses. The total energy released would then be 78 840 MJ and this must equal 63.2% of the initial available energy stored for the present analysis to be applied. Thus the total available heat stored

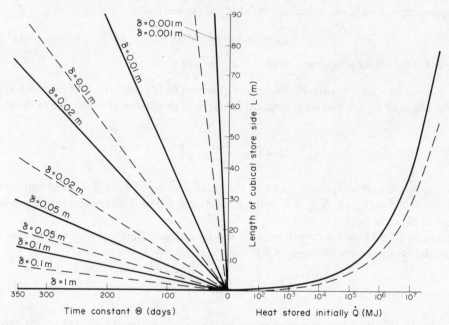

Fig. 11.7. General chart for sizing thermal stores and selecting insulation levels: ———— for sensible heat storage; ———— for latent heat storage.

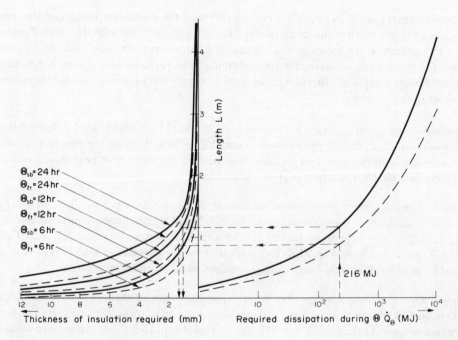

Fig. 11.8. Design chart for diurnal heat stores; ———— for sensible heat storage; ———— for latent heat storage.

equals 124 700 MJ. Refererring to the design chart (Fig. 11.9) it is seen that a cube of water of side 8.4 m is required with an insulation thickness of 83 mm. Because the temperature of the sensible heat store falls continuously during the discharge period, the instantaneous rate of heat dissipation, for a given level of insulation, is initially much higher than the design mean heat loss rate and is lower than the required dissipation in the latter part of the discharge period. There are two ways in which this variation may be overcome:

(a) The system can be over-designed with spare capacity to supply the design heat loss at $\Delta t = \Theta_{sb}$. Because the excess heat dissipated for $\Delta t < \Theta_{sb}$ must be rejected, this method is wasteful.

(b) Alternatively, the store can be designed with variable boundary insulation. The minimum and maximum effective thicknesses of the insulant can then be calculated from the initial temperature and that after a time interval Θ_{sb}. The design rate of heat loss (see Appendix to this chapter) would be given by

$$Q_t = U(T_{t=0} - T_\infty)\exp(-\Delta t/\Theta_{sb}),\qquad(11.18)$$

where U is a function of insulant effective thickness. In practice the rates of heat addition to or subtraction from a thermal store would be variable and so both input and exhaust facilities must have built-in controls.

11.8.1.2   LATENT HEAT ACCUMULATORS

The effectiveness of a latent heat store depends upon the latent heats of melting and solidification of its constituent materials. The phase change store has a much higher inherent capacity because much more energy is involved in phase transition than in increasing sensible heat

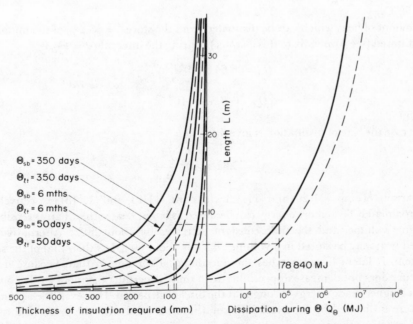

Fig. 11.9. Design chart for long-term thermal stores: ———— for sensible heat storage; ————— for latent heat storage.

alone. A further advantage is that, there being no temperature change involved, heat is transferred to the system in steady state, minimising losses in efficiency due to transient effects. The latent-heat store also operates as a thermal rectifier at constant temperature.

An applicable time constant for latent heat stores which allows for partial phase change may be defined (see Appendix to this chapter) as

$$\Theta_{lt} = 0.632 \frac{\rho V H_{fg}}{UA\Delta T},$$ (11.19)

where H_{fg} is the latent heat of fusion of the storage material and ΔT is now the fixed difference between the melting-point and the environmental temperature. Comparing this expression with that relating to sensible heat stores (eqn. (11.7)) it can be seen that $0.632 H_{fg}/\Delta T$ replaces c_p in that analysis. Table 11.5 shows that store volumes may be reduced by up to 50% if the store temperature is allowed to remain constant at temperatures other than 50°C (above a datum of 0°C) within the range zero to 100°C. Typically, $H_{fg}/\Delta T \simeq 7000$ J kg^{-1} K^{-1} and $\rho \simeq 1500$ kg m^{-3}. Substituting these values into eqn. (11.19) together with the U-value selected previously, then

$$\Theta_{lt} = 3.14 \times 10^7 L\delta,$$ (11.20)

Θ_{lt} being in seconds and L and δ in metres.

The heat stored initially (at 50°C for comparison with values for sensible heat storage in water) is given by

$$\dot{Q} = \rho V H_{fg}.$$ (11.21)

On substituting values,

$$\dot{Q} = 10.5 \times 10^6 L^3 \Delta T$$ (11.22)

or

$$\dot{Q} = 525 \times 10^6 L^3$$ (11.23)

if $\Delta T = 50$°C.

The amount of heat which can be dissipated from the store ($= 63.2\%$ of the initial available heat stored (see Appendix to this chapter)) during the interval $\Delta t = \Theta_{lt}$ is

$$\dot{Q}_{lt} = 6.64 \times 10^6 L^3 \Delta T$$ (11.24)

or

$$\dot{Q}_{\Theta_{lt}} = 3.32 \times 10^8 L^3$$ (11.25)

if $\Delta T = 50$°C.

In this case the rate of dissipation is invariant, i.e.

$$\dot{Q} = 3.32 \times 10^8 \frac{L^3}{\Theta_{lt}}.$$ (11.26)

Comparisons of eqns. (11.20) and (11.22) with eqns. (11.13) and (11.15) respectively show that approximately 60% more energy can be stored as latent heat rather than as sensible heat in the same volume, and the time constant increases by about 60%. Alternatively, the prescribed heat may be stored in a volume which is about 60% less than that of the sensible heat system. A latent heat store has the additional advantage in that, because its mean temperature does not decrease appreciably with time, it permits the more easily controlled removal of higher grade energy throughout the discharge period. However, the accumulator temperature is dictated by the availability of a suitable solid with the prescribed melting-point. In practice, therefore, the chosen example with a store temperature of 50°C is an abstraction from reality.

TABLE 11.5. MATERIALS FOR LATENT HEAT STORAGE

Material	Melting point (°C)	Heat of fusion H_{fg} (10^5 J kg^{-1})	Density ρ (kg m^{-3})	Volumetric heat of fusion H_{fg} (10^8 J m^{-3})	Temperature change for an equivalent mass of water (°C)	Temperature change for an equivalent volume of water (°C)	$H_{fg}/\Delta T^{(a)}$ (J kg^{-1} K^{-1})	Required store volume compared with water at 50°C (m^3)	Required store volume compared with water at the melting-point of the latent heat material (m^3)
Ice	0	3.36	913	3.07	80	73	—	—	—
Calcium chloride hexahydrate	29–39	1.75	1620	2.84	42	68	6000	0.67	0.43
Sodium carbonate decahydrate	32–36	2.70	1425	3.85	64	91	8000	0.53	0.36
Sodium sulphate decahydrate	32	2.34	1517	3.55	58	88	7300	0.60	0.38
Calcium nitrate tetrahydrate	41	2.10	1819	3.82	50	91	5100	0.55	0.45
Sodium	98	1.14	956	1.09	28	27	1160	1.92	3.76
Lithium	180	6.70	495	3.32	150	74	3720	0.63	2.26[b]
Sodium hydroxide	322	2.09	2130	4.45	69	147	649	0.47	3.02[b]
Aluminium	660	4.02	2600	10.50	96	250	609	0.20	2.63[b]
Sodium chloride	800	4.86	2119	10.30	116	246	607	0.20	3.24[b]

[a] Assuming that the surroundings are maintained at 0°C.
[b] As water boils at 100°C at atmospheric pressure the comparison is strictly invalid.

Latent heat storage system behaviour. Equations (11.20) and (11.22) describe the effects of accumulator dimensions and insulant thickness on the amount of heat to be stored and the period taken for a 63.2% reduction in the latent heat contained. The appropriate characteristics are presented in Figs. 11.7–11.9. In the examples chosen, for the same overall dissipation and for the same time constant, a cube of side 0.87 m with a 1.4 mm of insulant is required for the diurnal store, whereas a cube of side 6.0 m with 80 mm of insulant is needed for the 6 month store, i.e. the lengths of the sides are reduced compared with the sensible heat accumulator dimensions. Despite this superiority, the latent heat storage system will often be rejected in favour of a sensible heat system involving water because of its attendant convenience with respect to transportation and handling.

A guide has been developed which should aid the designer in the initial assessment of sensible and latent heat storage systems for short- and long-term applications. Insulant efficiency, stratification effects, and the transient and periodic behaviour of the storage medium and its auxiliaries should be considered in greater depth once the primary design parameters—the store capacity and the insulation level—have been chosen. Thus the values presented for insulant thicknesses should be taken as idealised minima, which must be increased appropriately to offset extraneous heat losses along feed pipes and mechanical supports.

11.8.1.3 THERMAL STORAGE IN SOLUTIONS

Nearly all salts have positive or negative heat of solutions, and solubility increases as the temperature is raised. If, therefore, a mixture of a salt which absorbs heat of solution and its saturated solution is heated, the resulting richer solution will absorb more energy. The apparent specific heat of the mixture will thus be greater than that of its components alone. The process is reversible on cooling, and, because the presence of the salt increases the boiling-point, energy may be stored at a higher temperature than possible in the solvent alone. The product of the heat of solution and the coefficient of solubility is a useful guide in the selection of suitable materials (Table 11.6). It is necessary to list values at different temperatures since many saturated solutions contain very much more salt than solvent, as a result of which the salt constitutes the main mass and volume of the mixture.

TABLE 11.6. SOLUTION HEAT CAPACITIES OF COMPOUNDS

Material	Heat of solution (MJ mol^{-1})	Solubility change (mol kg^{-1} K^{-1}) × 10^3	Solution heat capacity (kJ kg^{-1} K^{-1})	
			At 20°C	At 100°C
Ammonium nitrate	27.09	1.31	12.1	3.5
Potassium nitrate	36.13	0.27	7.4	2.8
Calcium nitrate hexahydrate	33.45	0.18	2.0	1.3
Sodium phosphate	54.43	0.08	3.7	2.0

11.8.1.4 THERMAL STORAGE USING REVERSE OSMOSIS

Osmosis occurs between solutions separated from their solvent by means of a semi-permeable membrane when the solvent diffuses preferentially into the solution. If this diffusion is resisted, osmotic pressures are set up across the membrane that may be of the order of many atmospheres. If a pressure in excess of the osmotic pressure is applied, the solution loses solvent and becomes more concentrated. Thus for a salt having a positive heat of solution, cooling will occur, or, under isothermal conditions, heat will be absorbed. When the pressure is removed the solvent passes back into the solution with the liberation of heat. A storage system based upon this principle has the advantage of the absence of any solid phases with the attendant thermal diffusivity limitations.

11.9   CONCLUSIONS

As the cost of high-grade energy rises its use for low-grade applications becomes prohibitive. Thus the employment of low-grade energy collection and storage systems will become more economic. Feasible systems should be cheap, simple to construct, and have reasonable working efficiencies. The "greenhouse" effect and the solar pond have been widely adopted because they obey these criteria. The success of the few installations in use highlight the growing need for immediate research and development in this hitherto neglected field.

APPENDIX
TIME CONSTANTS

The time constant for a sensible heat store. During the thermal discharge, the mean temperature and the instantaneous rate of heat dissipation change continuously. To simplify the analysis it is assumed that the temperature of the store T is a function of time t only, and is uniform throughout the system at any instant. It is also assumed that

rate of heat lost from the store = rate of heat transferred through the insulation,

i.e.,
$$- \rho c_p V \, dT = UA(T - T_\infty) \, dt. \tag{11.A1}$$

Integrating between $t = 0$ (corresponding to $T = T_o$) and $t = \Delta t$ gives

$$\log_e \left(\frac{T - T_\infty}{T_{t=0} - T_\infty} \right) = - \left(\frac{UA}{\rho c_p V} \right) \Delta t. \tag{11.A2}$$

The group of parameters $(\rho C_p V / UA)$ has the dimension of time and is defined as the "time constant" Θ_{sb}. When Δt is chosen to be equal to Θ_{sb}, then

$$\left(\frac{T - T_\infty}{T_{t=0} - T_\infty} \right) = e^{-1} = 0.368.$$

So, physically, Θ_{sb} is the time taken for the temperature excess $(T - T_\infty)$ of the system over its surroundings to fall to 36.8% of its initial value, which was $(T_{t=0} - T_\infty)$. Thus, at time Θ_{sb} the store temperature is given by

$$T_{\Theta_{sb}} = T_\infty + 0.368(T_{t=0} - T_\infty)$$

and, at any time Δt,

$$T = T_\infty + (T_{t=0} - T_\infty) \exp(-\Delta t / \Theta_{sb}). \tag{11.A3}$$

The instantaneous rate of heat dissipation can be calculated from

$$\dot{Q}_t = UA(T_{t=0} - T_\infty) \exp(-\Delta t / \Theta_{sb}) \tag{11.A4}$$

and the total amount of heat dissipated during Δt is obtained by integrating eqn. (11.A4), i.e.

$$\dot{Q}_{\Delta t} = \rho c_p V(T_{t=0} - T_\infty)[1 - \exp(-\Delta t / \Theta_{sb})]. \tag{11.A5}$$

Thus the total amount of heat dissipated in the interval $\Delta t = \Theta_{sb}$ is

$$\dot{Q}_{\Theta sb} = 0.632 \rho c_p V(T_{t=0} - T_\infty) \tag{11.A6}$$

or 63.2% of the initial heat stored.

The time constant for a latent heat store. Because the temperature of a latent heat store does not vary with time, the discharge rate is constant, and the latent heat store time constant cannot be expressed in terms of temperatures. Instead, Θ_{lt} is defined as the time taken for the heat content of the store to decrease to an amount equal to the heat content of an equivalent sensible heat store at time $t = \Theta_{lt}$, i.e. from eqn. (11.A6)

$$\dot{Q}_{\Theta lt} = 0.632 \, \rho V H_{fg} = UA(T_0 - T_\infty)\Theta_{lt}$$

or

$$\Theta_{lt} = 0.632 \left(\frac{\rho V H_{fg}}{UA\,\Delta T} \right). \tag{11.A7}$$

Chapter 12

Prospectus for the Future

12.1 WHERE SHOULD ENERGY BE SAVED?

Fossil-fuel stocks are limited, and present indications show that if current rates of energy consumption continues unabated the world will face energy starvation sometime during the twenty-first century. Unless life-styles prevalent in industrialised societies alter radically, or a new inexhaustible energy source is tapped, the cost of energy will continue to rise at an ever-increasing rate. There is no firm evidence that existing technologies will produce such a source in the near future. Thus there is no alternative to the widespread application of energy-conserving techniques.

Changes in marketing techniques and consumer-spending patterns could considerably reduce energy consumption in production. This would involve an end to "built-in obsolescence", the manufacture of disposable items, unnecessary packaging, convenience products, and labour-saving devices and systems, and so would demand a general decline in what we have come to define as "living standards".

A breakdown of national energy consumption into various sectors shows that half the energy released in the United Kingdom is dissipated during electricity generation and for the heating of buildings. Thirty per cent of the total energy is rejected from power stations in warm water at about 70°C, whilst another 30% is required for space heating rooms at about 25°C. Thus if the waste heat from electricity production was directed to heat buildings, 30% of the national fuel bill would disappear.

A further reduction (up to $\sim 15\%$) could be achieved by the proper energy-conscious design and construction of buildings. This would involve:

(a) the selection of optimum inside environmental conditions;
(b) a satisfactory analysis of periodic diurnal and annual thermal loads;
(c) the sensible application of thermal insulation;
(d) the most efficient use of all energy flows available;
(e) the selection of the most energy-cost-effective controlled heating systems; together with
(f) the planned systematic maintenance of, not only the conditioning plant and equipment, but also the building and its insulating components.

Energy flows can be redirected, inhibited, or enhanced using solar energy collectors, thermal rectifiers, heat exchangers, heat pipes, heat pumps, thermal accumulators, and thermal isolation technology. Additional energy can be harnessed from the sun, the winds, the tides, the waves, and the earth.

12.2 **TOTAL ENERGY**

The production and distribution of electricity from fossil fuels is an extremely wasteful process, more than 60% of the heat content of the fuel being discarded in cooling towers whilst transmission losses account for a further 8%. Electricity is the purest and most convenient form of energy being capable of conversion at almost 100% efficiency to other forms of energy.

Electricity involves the chain of reactions:

$$\text{fuel} \rightarrow \text{heat} \rightarrow \text{mechanical energy} \rightarrow \text{electrical energy}.$$

The most inefficient link in this chain is in the conversion from heat to mechanical energy. This is limited by the Carnot relationship

$$\eta < \frac{T_2 - T_1}{T_1},\tag{12.1}$$

where T_2 and T_1 are the highest and lowest temperatures attainable in the system. In practice, T_2 represents the maximum fluid temperature achieved inside the heat engine, whilst T_1 is limited by the lowest exhaust temperature allowable:

i.e. for a gas turbine: $T_1 \simeq 770 \text{ K}; \quad T_2 \simeq 1070 \text{ K},$

and so $\eta_{max} = 28\%,$

and for a steam turbine: $T_1 \simeq 300 \text{ K}; \quad T_2 \simeq 850 \text{ K},$

and so $\eta_{max} = 65\%.$

(In practice most steam plants operate at $\sim 37\%$ efficiency.)

Total energy operation [80] seeks to obtain electricity in a way that uses nearly all the energy contained in the fuel instead of only a small fraction. It generally describes schemes to generate shaft power for electricity generation combined with recovered heat using a thermal prime mover. There are three basic methods by which this may be achieved. These alternatives are listed below, classified in order of scale.

(a) *Large-scale total energy installations.* Large existing steam-operated power stations may be operated on what is known as the "intermediate take-off condensing" principle [80]. Steam is bled off along the turbine body at temperatures (90–180°C) higher than the normal exhaust temperature (~ 70°C). This may be used directly or to heat water. The resulting hot fluids can be used for district heating or air conditioning in nearby urban areas. Unfortunately, most recently built power stations are situated in rural areas. Nevertheless, alternative uses for this low-grade energy exist in agriculture, desalination, local industries, and trading estates.

(b) *Intermediate scale installations.* If electricity generation policies abandoned the grid system, small power plants could be built as part of a housing or industrial complex. Small steam-driven plants yield hot water at ~ 100°C. Alternatively, diesel engines, gas turbines, free piston engines, or stationary aircraft engines could be employed and the waste gases utilised.

(c) *Small-scale application.* Petrol-driven generators, or natural gas-fired fuel cells (under development [80]) may be used to produce electricity and low-grade heat for single dwellings.

Medium or small-scale total energy operations have the advantages that the employment of the waste heat increases overall efficiencies, electricity transmission losses are eliminated, and systems are able to survive independently from the grid system when national breakdowns occur. The cost of transporting hot water is currently about half the cost of transmitting electricity [80]. Decisions to install total energy systems are, however, inhibited by many factors. These include high capital costs, the need for standby facilities, the requirement for skilled operating and maintenance personnel, the space needed for the system, and the additional problems arising from the production of noise, vibration, and smoke. Electricity production is less efficient in smaller-scale plants, and fuel costs are higher than those borne by larger scale operations.

In a successful total energy installation, the equipment must be employed as fully as possible. When heating and cooling loads are subtracted from overall demands, much more even electricity consumption rates result. In order to maintain total efficiency, however, the waste heat must be utilised at a constant rate throughout the year. This is difficult to arrange in the United Kingdom when the system is designed predominantly for winter heating. Other uses for the waste heat in the summer months are desirable. These may include drying equipment, absorption air conditioning, the direct use of hot water, manufacturing processes, laundrywork, greenhouse heating, or soil warming.

12.3 ENERGY MANAGEMENT

The function of energy management is to monitor, record, analyse, critically examine, redirect, and control energy flows through systems so that energy is utilised with maximum efficiency [81].

Required system levels of performance should be clearly defined, scrutinised, and criticised. Designs for products, systems, processes, and procedures should ensure that implementation can be carried out with the least possible overall use of energy. The selection of plant and equipment should be based upon the most economic arrangements possible in energy terms for a given level of performance. Each process and procedure should be (a) optimised in relation to its required function, (b) installed, operated, and maintained correctly, and (c) supervised by trained personnel. Unnecessary energy wastage or premature rejection should be minimised.

Energy managers must be trained in the pertinent energy flow examination and modification techniques and should be supplied with the data, systematic procedural methods, and the software necessary to facilitate the application of energy thrift. These techniques already exist in part and are being used by method study engineers to achieve the most productive usage of manpower and materials.

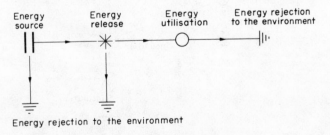

Fig. 12.1. Basic energy flow sequence.

12.3.1 **Energy flow charts**

An energy flow chart sets out the sequence and modes of energy flows through systems by recording—superimposed over a plan of a mechanical structure—all stations of energy release, utilisation, storage, and rejection to the environment. Some of the appropriate symbols for these operations already exist in method study software. Thus the following legend is proposed:

Energy event	Symbol	Classifications
Storage	─╫─	
Release	✳	
Utilisation	○	Primary operations
Rejection to the environment	⊥̲	
Conversion	⊗	
Heat exchange	⊗	Secondary operations
Flow resistance	─Ⱳ─	

A basic flow sequence is shown in Fig. 12.1. Conversion or heat transfer may be introduced between any two of the fundamental operations shown, and resistances to energy flow may be inserted anywhere in the chain of energy events. Thus an improved sequence is described in Fig. 12.2.

Modes of energy include thermal, electrical, chemical, mechanical, and acoustical forms. The nature may be indicated on the flow chart using appropriate colours or subscripts. Maximum, mean, and minimum amounts of energy in transit or storage and their modes of transportation can be indicated on lines connecting operations.

Energy flow charts can illustrate steady-state or transient energy flows through established structural systems, the historical or projected energy contents of products, or the energy flows in on-going continuous processes. The system examined can comprise a whole system or process or can be used to examine a single component of the whole.

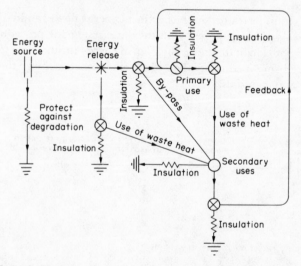

Fig. 12.2. Optimised basic energy flow sequence for the most efficient use of energy.

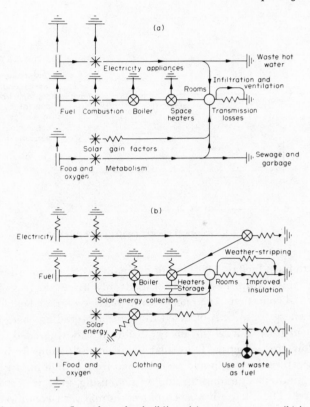

Fig. 12.3. Steady-state energy flows through a building: (a) common system; (b) improved system.

Examples of the use of energy flow charts. Figures 12.3–12.5 demonstrate the construction of energy flow charts for the following simple situations:

(a) steady-state energy flow through a building (Fig. 12.3);
(b) the historical energy content of a product (Fig. 12.4);
(c) energy flows through a factory (Fig. 12.5).

The flow diagram aids comprehension of overall system activity. Energy conservation sense is, in general, common sense. As in the application of method study techniques, however, the overall picture is so complex that—unless systematic procedures are available for observation, recording, and critical assessment—obvious modifications required are difficult to pinpoint.

The following general rules for practical energy management emerge from studies of flow diagrams:

(a) High-grade thermal energy has a temperature appreciably different from that of the environment, and thus may be "hot" or "cold" with respect to the environmental conditions.
(b) All energy is down-graded to environmental conditions during utilisation.
(c) All energy activity involves heat rejection.
(d) If additional thermal energy is to be added to a down-grading energy flow system it is more efficient to add low-grade energy at the lower end of the energy chain. Thus it is

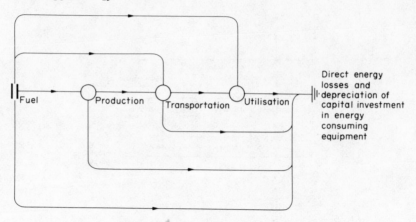

Fig. 12.4. Historical energy content of a product.

ridiculous to burn fuel at 2000°C or to dissipate electricity in order to warm a room where the temperature needs to be raised only a few degrees above the outside environmental conditions. It is energy sense to use low-grade heat for this purpose.

(e) If thermal energy is to be subtracted from a down-grading energy flow system it is more efficient to withdraw this energy before it is downgraded, using an external source of low-grade energy. Thus solar gains or dissipations from lighting or other high-temperature devices should be extracted from a building by cooling windows, louvres or shutters, luminaries, etc., *in situ*, using inlet water at environmental temperatures. The alternative, as employed in many air-conditioning systems, is to allow the energy to degrade to inside environmental conditions, where it is at last withdrawn from the flow system using high-grade chilled water or refrigerant [82, 83].

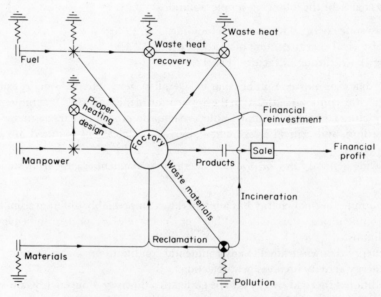

Fig. 12.5. Optimised energy flows through a factory.

(f) High-grade energy (i.e. to be regarded as the energy content of substances which are maintained at temperatures greater than $\pm 15°C$ from the environmental temperature) should not be allowed to dissipate directly to the environment. The energy rejected from an activity should be collected and stored to be used at another time, or should be redirected for a high-grade purpose using a heat pump, or for a low-grade purpose using a heat exchanger, at some other energy demand point.

12.3.2 Critical assessment of energy usage

Once the flow chart has been established, problem areas can easily be identified. Each operation must be critically examined with a view to eliminating or reducing local and overall energy consumption. Part of the energy flow may be recycled by (a) feedback between stages, (b) partial bypass of an operation, or (c) redirecting rejected energy to a lower-grade useful purpose. A suitable critical examination chart is reproduced in Table 12.1. String diagrams could prove useful for rapid modifications, and computer-based operational research techniques can be used to analyse the overall effects of suggested changes and to optimise the system in terms of overall energy consumption. Appraisals and recommendations should include flow charts of the existing and proposed systems, present and proposed overall energy costs, capital expenditure required for modifications, and a time-scale for implementation.

TABLE 12.1. CRITICAL ASSESSMENT SHEET

	Description of operation			
	Primary question	Present facts	Alternatives	Selected alternative
Purpose	What is done?	Is it necessary? Why?	What else could be done?	What should be done?
Means	How is it done?	Why like this?	How else could it be done?	How should it be done?
Place	When is this done in the energy flow?	Why then?	Where else could it be done?	Where should it be done?
Source Grade	Where does the input energy come from?	Why this source?	What other source could be used?	What other source should be used?
Sink Grade	Where does the rejected energy go?	Why is it allowed to go there?	Where could it be redirected?	Where should it be redirected?

12.3.3 Essential qualities of an energy manager

Whereas the work study engineer must have a solid grounding in production techniques, the manager is a rarer breed, requiring expertise in, not only systems analysis, but also in thermodynamics, heat and mass transfer, fluid flow, energy technology, and related disciplines.

12.4   ENERGY ACCOUNTING

An energy audit [3] of an existing system or flow process records local and overall energy running costs. A component or product can also be examined in terms of its historical energy content. An energy budget forecasts the total amount of energy required to establish and sustain a planned operation. Energy accounts list capital and running costs of systems using energy instead of cash as an economic base. Such procedures, although often difficult to implement, aid the effective utilisation of energy resources. Energy costs are often obscured in cash-based paperwork, yet, as the price of energy rises, these will become appreciable factors influencing management decisions and selections between options. Energy-based accounts may be more durable than those founded upon ever-changing and inflating monetary systems. Thus optimum projected procedures deduced from energy considerations are less likely to alter in the long term.

12.5   ENERGY CONSERVATION IN PRACTICE

The estimates of remaining fossil-fuel reserves presented in Chapter 1 predict a rather gloomy future. The prospects are, however, not as inhibiting as they appear. The profligate consumption of energy which has taken place since the industrial revolution has been founded upon the availability of cheap and plentiful fossil fuels. The thoughtless exploitation of these fuels has caused devastating environmental pollution. The major breakthrough in energy and environmental conservation will occur when it is universally recognised that resources are limited; energy will then be treated as a precious commodity to be utilised in the most efficient manner with the minimum deleterious impact on the environment. Energy auditors will be appointed; energy accounts will be kept; energy flows will be closely monitored; energy-consuming operations will be improved; and new techniques and procedures will be developed to increase energy effectiveness. People are already becoming more energy conscious: the rate of use of fossil fuel in the United Kingdom has fallen since 1974.

The energy shortage strikes worldwide. All nations must conserve. Britain is an industrial nation which relies heavily on the import of fuels to feed manufacturing industries. Because of this reliance, the oil crisis affected the United Kingdom to a greater extent than many other countries. In the immediate future, North Sea oil and gas may be able to supply all internal needs for a period of up to 25 years. The danger is that during this era of relative self-sufficiency Britain may forget the energy crisis whilst others conserve. Then, when all the indigenous fuel is used up, we will be left with outmoded energy technologies, poorly insulated buildings, and energy-hungry machines and production processes.

Appendix I

Solution of Linear Equations

A set of linear equations describing a closed system may be represented in general algebraic form as

$$\left.\begin{array}{l} a_{11}x_1 + a_{12}x_2 + a_{13}x_3 + \ldots + a_{1n}x_n = b_1 \\ a_{21}x_1 + a_{22}x_2 + a_{23}x_3 + \ldots + a_{2n}x_n = b_2 \\ a_{31}x_1 + a_{32}x_2 + a_{33}x_3 + \ldots + a_{3n}x_n = b_3 \\ \qquad\qquad \vdots \\ a_{n1}x_1 + a_{n2}x_2 + a_{n3}x_3 + \ldots + a_{nn}x_n = b_n \end{array}\right\} \tag{A.I.1}$$

or, in matrix format, as

$$\begin{bmatrix} a_{11} & a_{12} & a_{13} & \cdots & a_{1n} \\ a_{21} & a_{22} & a_{23} & \cdots & a_{2n} \\ a_{31} & a_{32} & a_{33} & \cdots & a_{3n} \\ \vdots & & & & \\ a_{n1} & a_{n2} & a_{n3} & \cdots & a_{nn} \end{bmatrix} \times \begin{Bmatrix} x_1 \\ x_2 \\ x_3 \\ \vdots \\ x_n \end{Bmatrix} = \begin{Bmatrix} b_1 \\ b_2 \\ b_3 \\ \vdots \\ b_n \end{Bmatrix} \tag{A.I.2}$$

or, simply, as

$$[A]\{x\} = \{b\}, \tag{A.I.3}$$

where $[A]$ is a matrix of coefficients and $\{x\}$ and $\{b\}$ are vectors.

ELIMINATION METHOD OF SOLUTION

The elimination method is used in elementary algebra for solving systems of simultaneous linear equations; e.g. a simple example:

$$x_1 + 2x_2 = 3, \qquad \text{(i)}$$

$$2x_1 + x_2 = 5. \qquad \text{(ii)}$$

If eqn. (i) is multiplied by 2 and eqn. (ii) is subtracted from the modified version of eqn. (i),

$$x_2 = 0.333.$$

Substituting this value for x_2 in eqn. (ii),

$$x_1 = 2.333.$$

This technique can be extended to solve a set of n simultaneous linear equations (A.I.1) using the following procedure:

I. Divide the first equation by a_{11}.

II. Use the result to eliminate x_1 in all the other equations.

III. Divide the second equation by the modified coefficient of x_{22}.

IV. Use the result to eliminate x_2 in eqns. (A.I.3) to (n).

V. Continue until a modified matrix of coefficients is produced:

$$\begin{bmatrix} a'_{11} & a'_{12} & a'_{13} & \cdots & a'_{1n} \\ 0 & a'_{22} & a'_{23} & \cdots & a'_{2n} \\ 0 & 0 & s'_{33} & \cdots & a'_{3n} \\ & & & & \\ & & & & \\ & & & & a'_{nn} \end{bmatrix} \times \begin{Bmatrix} x_1 \\ x_2 \\ x_3 \\ \cdot \\ \cdot \\ x_n \end{Bmatrix} = \begin{Bmatrix} b'_1 \\ b'_2 \\ b'_3 \\ \cdot \\ \cdot \\ b_n \end{Bmatrix}$$

VI. x_n can then be obtained from the nth equation, i.e.

$$x_n = \frac{b'_n}{a'_{nn}}.$$

VII. This value for x_n is then used to solve equation $n - 1$ for x_{n-2}, etc., until all values for $x_1 \ldots x_n$ have been obtained.

Pivotal condensation [48] is often used to avoid the generation of excessively large numerical values for the coefficients during the elimination process. Columns and rows are interchanged to place the largest coefficient present in the a_{11} position. Before step III the largest remaining coefficient is placed in the a_{22} position and so on. Sometimes row interchanges alone are sufficient.

Interchanging rows and columns of matrix equations

EXAMPLES

Original matrix

$$\begin{bmatrix} a_{11} & a_{12} & a_{13} \\ a_{21} & a_{22} & a_{23} \\ a_{31} & a_{32} & a_{33} \end{bmatrix} \begin{Bmatrix} x_1 \\ x_2 \\ x_3 \end{Bmatrix} = \begin{Bmatrix} b_1 \\ b_2 \\ b_3 \end{Bmatrix}$$

Interchanging rows 1 and 2 of the original matrix

$$\begin{bmatrix} a_{21} & a_{22} & a_{23} \\ a_{11} & a_{12} & a_{13} \\ a_{31} & a_{32} & a_{33} \end{bmatrix} \begin{Bmatrix} x_1 \\ x_2 \\ x_3 \end{Bmatrix} = \begin{Bmatrix} b_2 \\ b_1 \\ b_3 \end{Bmatrix}$$

$$\underset{\text{change}}{\text{no}} \qquad \text{change}$$

Interchanging columns 1 and 2 of the original matrix

$$\begin{bmatrix} a_{12} & a_{11} & a_{13} \\ a_{22} & a_{21} & a_{23} \\ a_{32} & a_{31} & a_{33} \end{bmatrix} \begin{Bmatrix} x_2 \\ x_1 \\ x_3 \end{Bmatrix} = \begin{Bmatrix} b_1 \\ b_2 \\ b_3 \end{Bmatrix}$$

change   no
change

Interchanging rows 1 and 2 and columns 1 and 2 in the original matrix

$$\begin{bmatrix} a_{22} & a_{21} & a_{23} \\ a_{12} & a_{11} & a_{13} \\ a_{32} & a_{31} & a_{33} \end{bmatrix} \begin{Bmatrix} x_2 \\ x_1 \\ x_3 \end{Bmatrix} = \begin{Bmatrix} b_2 \\ b_1 \\ b_3 \end{Bmatrix}$$

change change

Rockey [50] further describes the manipulation of matrices, whilst Pennington [48] provides computer subroutines for digital solution by elimination. This method of solution is most suitable for steady-state systems.

GAUSS–SIEDEL METHOD OF SOLUTION

An alternative method for solving a system of linear equations is the iterative Gauss–Siedel method in which the eqns. (A.I.1) are rewritten so that, in each, all but one unknown is taken to the right of the equation; i.e. in general form:

$$\left. \begin{array}{l} a_{11}x_1 = b_1 - a_{12}x_2 - a_{13}x_3 \ldots - a_{1n}x_n \\ a_{22}x_2 = b_2 - a_{21}x_1 - a_{23}x_3 \ldots - a_{2n}x_n \\ a_{33}x_3 = b_3 - a_{31}x_1 - a_{32}x_2 \ldots - a_{3n}x_n \\ \quad . \qquad\qquad\qquad\qquad\qquad . \\ \quad . \qquad\qquad\qquad\qquad\qquad . \\ \quad . \qquad\qquad\qquad\qquad\qquad . \\ a_{nn}x_n = b_n - a_{n1}x_1 - a_{n2}x_2 \ldots - a_{n,n-1}x_{n-1} \end{array} \right\} \qquad \text{(A.I.4)}$$

A set of initial guessed values for $x_2, x_3, \ldots x_n$ is substituted into the right-hand side of the first equation to solve for x_1. This value, together with the initial estimates for $x_3 \ldots x_n$, are then substituted into the right-hand side of the second equation to solve for x_2. The initial value for x_2 is then discarded. The process continues through the equations until there is established a secondary set of values for $x_1 \ldots x_n$. The procedure is repeated until the system converges (i.e. successive values for each value of x do not differ by more than an assigned tolerance).

It is unfortunately not always absolutely certain that this process will converge. To increase the probability of success it is usual to arrange that the coefficients of the variables brought to the left-hand side of each equation have the largest absolute values in the row: e.g. simple example:

$$x_2 = \frac{3 - x_1}{2}, \qquad \text{(iii)}$$

$$x_1 = \frac{5 - x_2}{2}. \qquad \text{(iv)}$$

		x_1	x_2
Initial estimates		0	0
Substituting x_1 in (iii)		0	1.5
Substituting x_2 in (iv)		1.75	1.5
,,		1.75	1.625
,,		2.187	0.625
,,		2.187	0.406
,,		2.300	0.406
,,		2.300	0.350
,,		2.325	0.350
,,		2.325	0.337
		.	.
		.	.
		.	.
		2.333	0.333

The system is extremely amenable to digital computation [48] and may easily be extended to solve systems of non-linear equations.

Summary of procedure

I. Examine equations and determine which coefficient present has the greatest effect on the matrix.

II. Bring the equation containing this coefficient to the first row.

III. Bring the term containing this largest coefficient to the left-hand side of the first equation.

IV. Repeat steps I–III for the remaining equations until the entire system is rearranged.

V. Guess initial values.

VI. Solve the equations for the unknown values.

VII. Iterate until convergence occurs.

BASIC RELAXATION METHOD OF SOLUTION

The basic relaxation procedure is a relatively rapid tabular iterative method of solution used extensively prior to the era of easy access digital computers. The set of eqns. (A.I.1) is rewritten in the form:

$$\left.\begin{aligned}
a_{11}x_1 + a_{12}x_2 + \ldots + a_{1n}x_n - b_1 &= F_1 \\
a_{21}x_1 + a_{22}x_2 + \ldots + a_{2n}x_n - b_2 &= F_2 \\
a_{31}x_1 + a_{32}x_2 + \ldots + a_{3n}x_n - b_3 &= F_3 \\
& . \\
& . \\
& . \\
a_{n1}x_1 + a_{n2}x_2 + \ldots + a_{nn}x_n - b_n &= F_n
\end{aligned}\right\} \quad \text{(A.I.5)}$$

where $F_1 \ldots F_n$ are called "residuals". The object of the relaxation method is to reduce the values of the residuals to as near zero as possible. Initial values for $x_1 \ldots x_n$ and $b_1 \ldots b_n$ are first estimated. An "operations" table is then set up which shows the effects of unit positive increments in the variations of $x_1 \ldots x_n$ and $b_1 \ldots b_n$ on the residuals. The system is then "relaxed" by altering the values of $x_1 \ldots x_n$ and $b_1 \ldots b_n$ until the value of the lowest residual present is less than a prescribed allowable tolerance: i.e. simple example:

$$4x_1 - x_2 = 56, \qquad \text{(v)}$$
$$-x_1 + 2x_2 = 34. \qquad \text{(vi)}$$
$$F_1 = -4x_1 + x_2 + 56.$$
$$F_2 = x_1 - 2x_2 + 34.$$

OPERATIONS TABLE

	ΔF_1	ΔF_2
$\Delta_{x1} = 1$	-4	1
$\Delta_{x2} = 1$	1	-2
Block $\Delta_{x1} = 1$ and $\Delta_{x2} = 1$	-3	-1

Values of residuals

Initial estimates	$x_1 = 0$	$x_2 = 0$	$F_1 = 56$	$F_2 = 34$	The procedure is to
$\Delta x_1 = 14$	$x_1 = 14$	$x_2 = 0$	$F_1 = 0$	$F_2 = 48$	reduce the value of
$\Delta x_2 = 24$	$x_1 = 14$	$x_2 = 24$	$F_1 = 24$	$F_2 = 0$	the currently
$\Delta x_1 = 6$	$x_1 = 20$	$x_2 = 24$	$F_1 = 0$	$F_2 = 6$	largest residual
$\Delta x_2 = 3$	$x_1 = 20$	$x_2 = 27$	$F_1 = 3$	$F_2 = 0$	to zero
$\Delta x_1 = 1$	$x_1 = 21$	$x_2 = 27$	$F_1 = -1$	$F_2 = 1$	
$\Delta x_1 = -0.3$	$x_1 = 20.7$	$x_2 = 27$	$F_1 = 0.2$	$F_2 = 0.7$	
$\Delta x_2 = 0.4$	$x_1 = 20.7$	$x_2 = 27.4$	$F_1 = 0.6$	$F_2 = -0.1$	
$\Delta x_1 = 0.1$	$x_1 = 20.8$	$x_2 = 27.4$	$F_1 = 0.2$	$F_2 = 0.0$	

The values are seen to be converging. The process may be continued for further accuracy.

Short cuts

The amount of computation time required for the relaxation technique may often be reduced if the following short cuts are adopted when possible:

(i) The physics of the problem should be studied to make good initial estimates of the variables.

(ii) When evaluating potential fields, a coarse grid of nodal points should be used for preliminary calculations. The results of this rough initial study can then be used as the initial estimates in the analyses of finer grids.

(iii) *Over-relaxation.* Instead of reducing the residuals to zero at each step, larger increments of the variables can be used so that a positive value for a residual becomes negative and *vice versa.* This can considerably speed up the process of convergence, although care must be taken to prevent the process becoming unstable (i.e. if the increment in the variable is too large a positive value for a residual will become a *larger* negative value).

(iv) *Block operation.* Sometimes the process may be accelerated by changing each of the variables by the same amount. (N.B.—The operation in which all values for $x_1 \ldots x_n$ are altered by unity is sometimes known as "unit block operation" [18]).

(v) Once again, to avoid divergence from the set of values sought, it is advisable to select the variable to be altered which has the greatest effect on the largest residual present.

Appendix II

International system of Units (SI)

SI units (Système International d'Unités) [84] have been used throughout the text. This system was recommended by the International Organisation for Standardisation in 1960 and has since been adopted by many countries. The SI system is superior to others in current use because it is a completely "coherent" system, the product or quotient of any two quantities leading to a unit of the resultant quantity with no need for multiplying factors. As well as the benefits inherent in uniformity, the system has the additional advantage that analogies between different processes are not obscured by the use of different units. Thus, for example, amounts of energy associated with thermal, electrical, chemical, mechanical, and other processes may be expressed in terms of a common unit: either the Joule or the Watt, depending upon whether the energy is "stationary" or in transit.

The International System of Units is based upon six basic units and two supplementary units listed in Table A.II.1. Each other unit is built up from the appropriate basic units, and so may be expressed in terms of the basic units, even though some special names and symbols are also recommended for some of the more important derived units (Table A.II.2).

TABLE A.II.1. BASIC AND SUPPLEMENTARY SI UNITS

Quantity	Name of unit	Symbol
Mass	kilogram	kg
Length	metre	m
Time	second	s
Thermodynamic temperature	degree Kelvin	K
Electric current	ampere	A
Luminous intensity	candela	cd
Plane angle	radian	rad
Solid angle	steradian	sr

1 gram: mass of 1 cc of water at 0 °C.

1 metre: originally one ten-millionth part of the distance from the north pole to the equator through Paris, France. Redefined in 1960 as the length equal to 1 650 763.73 wavelengths *in vacuo* of the radiation corresponding to the transition between the levels $2p_{10}$ and $5d_5$ of the isotope 86/36 Kr.

210

TABLE A.II.2. DERIVED SI UNITS

Quantity	Name of unit	Symbol	Basic equivalent units
Force	Newton	N[a]	$kg\ m\ s^{-2}$
Work, energy, quantity of heat	Joule	J	$N\ m \equiv kg\ m^2\ s^{-2}$
Power	Watt	W	$J\ s^{-1} \equiv kg\ m^2\ s^{-3}$
Pressure	Pascal	Pa	$N\ m^{-2} \equiv kg\ m^{-1}\ s^{-2}$[b]

[a]The weight of a mass of M kg is a force of Mg Newtons, where g is the local value of the acceleration due to gravity.

[b]1 bar $\equiv 10^5\ N\ m^{-2}$.

Because some SI units are of inconvenient size, multiplying prefixes are available (Table A.II.3). These prefixes are printed immediately adjacent to the unit symbols with which they are associated; they then become part of the symbol, i.e.

$$1\ cm^2 \equiv (10^{-2}\ m)^2$$

$$1\ MN\ m^{-2} \equiv M(N\ m^{-2})$$

$$1\ GJ\ m^{-3} \equiv 10^9\ J\ m^{-3},\ etc.$$

SI units of all other quantities used in the text are given in the nomenclature.

TABLE A.II.3. RECOMMENDED MULTIPLYING PREFIXES

Value	Name	Symbol
10^{12}	tera	T
10^9	giga	G
10^6	mega	M
10^3	kilo	k
10^{-3}	milli	m
10^{-6}	micro	μ
10^{-12}	pico	p

Appendix III

Dimensionless Groups

Quantity	Physical significance	Symbol	Expression
Biot modulus	$\dfrac{\text{Internal resistance}}{\text{External resistance}}$	Bi	hL/k_s
Fourier modulus	Time dependence	Fo	$\alpha t/L^3$
Grashof number	$\dfrac{\text{Buoyancy forces}}{\text{Viscous forces}}$	Gr	$\dfrac{\rho^2 g\beta\Delta TL^3}{\mu^2}$
Nusselt number	$\dfrac{\text{Conductive resistance}}{\text{Convective resistance}}$	Nu	hL/k_f
Prandtl number	$\dfrac{\text{Viscous transfer}}{\text{Heat transfer}}$	Pr	$\nu/\alpha = \mu c_p/k_f$
Reynolds number	$\dfrac{\text{Inertia forces}}{\text{Viscous forces}}$	Re	$\dfrac{\rho uL}{\mu} = \dfrac{uL}{\nu} = \dfrac{GL}{\mu}$

Combinations of primary dimensionless groups

Peclet number $\qquad Pe = Re \times Pr$

Stanton number $\qquad St = Nu/Pe = Nu/Re \times Pr = h/\rho u c_p$

Colburn "j" factor $\qquad j = St \times Pr^{2/3}$

N.B. Subscripts: f, of fluid; s, of solid.

Appendix IV

Properties of Selected Materials

(a) Properties of some commonly used metals at 20 °C

	Density ρ (kg m^{-3})	Thermal conductivity k (W m^{-1} K^{-1})	Specific heat c_p (J kg^{-1} K^{-1})	Volumetric heat capacity ρc_p (10^6 J m^{-3} K^{-1})	Thermal diffusivity α (10^{-6} m^2 s^{-1})
Aluminium	2790	164	883	2.46	66.6
Brass	8520	111	385	3.28	33.8
Cast iron (4% C)	7270	52	419	3.05	17.6
Copper	8960	386	389	3.48	110.9
Magnetite (Fe$_2$O$_3$)	5177	1.9	752	3.85	0.5
Steel (1% C)	7800	43	473	3.69	11.7

(b) Average properties of some common non-metallic solids at 20 °C

Ash	720	0.10	—	—	—
Brick (dry)	1785	0.45	837	1.49	0.30
(9% moisture)	1892	0.80	837	1.49	0.54
Cardboard (corregated)	105	0.047	—	—	—
Cement	1700	0.80	—	—	—
Clay	1458	1.28	879	1.28	1.00
Concrete	2110	1.10	897	1.89	0.62
(Stone)	2304	0.93	837	1.93	0.48
(10% moisture)	2240	1.211	837	1.87	0.64
Cotton	80	0.059	1300	0.10	0.59
Cotton wool	80	0.041	—	—	—
Corkboard	160	0.043	2000	0.32	0.13
Diatomaceous earth	320	0.062	879	0.28	0.22
Earth (coarse gravelly)	2050	0.521	1840	3.77	0.14
Fibre insulating board	237	0.048	—	—	—
Glass plate	2710	0.762	837	2.27	0.34
Glass wool	200	0.04	670	0.13	0.31
Granite	—	2.80	—	—	—
Ice (at 0 °C)	913	2.21	1930	1.76	1.26
Kapok	20	0.035	—	—	—
Marble	2600	2.77	808	2.10	1.32
Mineral wool	150	0.038	—	—	—
Mud	1840	0.43	—	—	—
Paper	—	0.1275	—	—	—
Plaster	400	0.10	—	—	—
Polystyrene (expanded)	25	0.034	—	—	—
Rockwool	128	0.029	—	—	—
Roofing felt	960	0.19	—	—	—
Sand (dry)	1520	0.346	—	—	—
(10% moisture)	1600	0.38	—	—	—

(b) Average properties of some common non-metallic solids at 20C° (*Cont.*)

	Density ρ (kg m^{-3})	Thermal conductivity k (W m^{-1} K^{-1})	Specific heat c_p (J kg^{-1} K^{-1})	Volumetric heat capacity ρc_p (10^6 J m^{-3} K^{-1})	Thermal diffusivity α (10^{-6} m^2 s^{-1})
Sandstone	2200	1.85	712	1.56	1.19
Sawdust	192	0.051	—	—	—
Slate	—	1.462	—	—	—
Snow (0°C)	555	0.459	—	—	—
Soil (dry)	—	0.85	1841	—	—
(wet)	—	0.64	—	—	—
Thatch	240	0.07	—	—	—
Urea formaldehyde foam	8	0.038	—	—	—
Wallboard	236.8	0.0476	—	—	—
Wood (oak)	700	0.19	239	0.17	1.12
(fir)	419	0.188	272	0.11	1.28
Wool	110.4	0.0357	—	—	—

(c) Properties of water at 0.1013 MN m⁻²

Temperature T (°C)	Density ρ (kg m⁻³)	Thermal conductivity k (W m⁻¹ K⁻¹)	Specific heat c_p (J kg⁻¹ K⁻¹)	Volumetric heat capacity ρc_p (10⁶ J m⁻³ K⁻¹)	Thermal diffusivity α (10⁻⁶ m² s⁻¹)	Dynamic viscosity μ (10⁻³ kg m¹ s⁻¹)	Kinematic viscosity ν (10⁻⁶ m¹ s⁻¹)	Coefficient of volumetric expansion β (10⁻³ K⁻¹)	Prandtl number Pr	Saturation pressure p (kN m⁻²)
Saturated liquid										
0	1000	0.569	4217	4.217	0.134	1.755	1.755	−0.06	13.02	—
20	998	0.603	4182	4.173	0.145	1.002	1.004	0.207	6.95	—
40	992	0.632	4179	4.146	0.152	0.651	0.656	0.385	4.31	—
60	983	0.653	4185	4.113	0.159	0.462	0.476	0.523	2.96	—
80	972	0.670	4197	4.079	0.164	0.350	0.360	0.641	2.19	—
100	958	0.681	4206	4.029	0.169	0.278	0.290	0.750	1.72	—
Saturated vapour										
0	0.00485	0.0173	1854	8.99	1900	0.0088	1800	0.0036	0.942	0.611
20	0.0173	0.0191	1866	32.28	590	0.0094	540	0.0034	0.918	2.34
40	0.0513	0.0204	1885	96.70	210	0.0101	196	0.0032	0.930	7.38
60	0.1300	0.0217	1915	248.95	87	0.0107	82	0.0030	0.947	19.92
80	0.2930	0.0231	1962	574.87	40	0.0114	38	0.0028	0.966	47.36
100	0.5980	0.0249	2028	1212.74	21	0.0121	20.2	0.0027	0.986	101.30

(d) Properties of dry air at 0.1013 MN m⁻²

T (°C)	ρ	k	c_p	ρc_p	α	μ	ν	β	Pr	p
−20	1.400	0.0223	1003.1	1404	15.8	0.0160	11.4	0.0039	0.720	000.000
−10	1.341	0.0213	1003.4	1345	15.8	0.0165	12.6	0.0038	0.716	000.000
0	1.286	0.0242	1003.8	1291	18.7	0.0170	14.9	0.0036	0.713	000.000
20	1.203	0.0250	1004.4	1208	20.6	0.0180	16.8	0.0034	0.710	000.000
40	1.130	0.0272	1005.6	1136	23.9	0.0190	18.4	0.0032	0.704	000.000
60	1.088	0.0290	1007.0	1096	26.4	0.0200	21.0	0.0030	0.697	000.000
80	1.000	0.0300	1008.0	1008	29.7	0.0210	23.3	0.0028	0.696	000.000
100	0.942	0.0318	1010.0	951	33.4	0.0220	23.35	0.0027	0.692	000.000

Appendix V

Useful values in SI units

Mean molecular weight of dry air $M_a = 28.97$ kg.

Mean molecular weight of steam $M_v = 18.02$ kg.

Density of dry air at 15.5°C, $\phi = 60\%$, $p = 10^5$ N m^{-2}, $\rho_a = 1.22$ kg m^{-3}.

Density of liquid water at 15.5°C $\rho_{wat} = 1000$ kg m^{-3}.

Universal gas constant $\mathbf{R}_{gas} = MR = 8.3143$ kJ kmol^{-1} K^{-1}.

Volume of one mol of the permanent gases (at 1.01325 bar and 0°C) = 22.4136 m^3.

Characteristic gas constant for dry air $\mathbf{R}_a = 287$ J kg^{-1} K^{-1}.

Characteristic gas constant for steam $\mathbf{R}_v = 462$ J kg^{-1} K^{-1}.

Absolute values of enthalpy are related to a datum of 273.15 K.

Mean specific heats at room temperatures

For air at constant pressure $c_{pa} = 1005$ J kg^{-1} K^{-1}.

For air at constant volume $c_{va} = 718$ J kg^{-1} K^{-1}.

For steam at constant pressure $c_{pv} = 4210$ J kg^{-1} K^{-1}.

For steam at constant volume $c_{vv} = 1810$ J kg^{-1} K^{-1}.

Adiabatic index for air at room temperatures and pressures = 1.4.

Latent heat of steam at 0°C = 2 500 000 J kg^{-1} K^{-1}.

References

1. I. G. C. Dryden, *The Efficient Use of Energy*, IPC Science and Technology Press, Department of Energy, UK, 1975.
2. P. Hill, and R. Vielvoye, *Energy in Crisis*, Robert Yeatman Ltd., London, 1974.
3. J. Lenihan and W. W. Fletcher, *Energy Resources and Environment*, vol. 1, Blackie, London, 1975.
4. G. F. C. Rogers, The energy perspective, *Chem. metall. Engng., Instn. Mech. Engrs.*, 61–65, Oct. 1974.
5. B. Crossland, The mechanical engineer and the environmental problems, *Chem. metall. Engng., Instn. Mech. Engrs.*, 117–121, Sept. 1975.
6. B. Commoner, *The Closing Circle*, Jonathan Cape, London, 1972.
7. G. F. C. Rogers, Energy conservation: choice or necessity?, *Chem. metall. Engng., Instn. Mech. Engrs.*, 65–69, May, 1975.
8. B. Ward and R. Dubos, *Only One Earth*, Andre Deutsch, 1972.
9. A. Tucker, The last judgement, *The Guardian*, 1 March 1976.
10. DTI, *Digest of United Kingdom Energy Statistics*, Department of Trade and Industry, HMSO, London, 1972.
11. J. C. Fisher, *Energy Crisis in Perspective*, John Wiley, London, 1974.
12. CPRS, *Energy Conservation*, Central Policy Review Staff, HMSO, London, 1974.
13. A. H. Corbett, Energy and the profession of engineering, *J. Instn. Engrs. Aust.* 4–16, May 1974.
14. P. T. Hinde and S. D. Probert, Thrift with thermal energy, *New Scient.* **65**, 210–211 (1975).
15. G. F. C. Rogers, Social consequences of an energy limited economy, *Chem. metall. Engng., Instn. Mech. Engrs.*, 59–62, Oct. 1975.
16. V. L. Streeter, *Fluid Mechanics*, McGraw-Hill, 4th edn., 1966.
17. J. G. Knudsen and D. L. Katz, *Fluid Dynamics and Heat Transfer*, McGraw-Hill, 1958.
18. F. Kreith, *Principles of Heat Transfer*, International Textbook Co., Scranton, Pennsylvania, USA, 1965.
19. IHVE, *Guide Books*, Institution of Heating and Ventilating Engineers, London, 1970.
20. Electricity Council, *Packaged Air Conditioning*, 1974.
21. W. H. McAdams, *Heat Transmission*, McGraw-Hill, 3rd edn., 1958.
22. P. W. O'Callaghan and S. D. Probert, Micro-topographic influence of nominally flat metallic surfaces upon their thermally reflective behaviour, *J. mech. Engng. Sci. Instn. Mrch. Engrs.* **19** (2) 65–75 (1977).
23. P. W. O'Callaghan, The effect of water vapour on the heat transfer performance of an air-to-air heat exchanger, MSc thesis, University of Wales, 1968.
24. ASHRAE, *Handbook of Fundamentals*, American Society of Heating, Refrigerating, and Air-conditioning Engineers, 1972.
25. I. D. Griffiths and P. R. Boyce, Performance and thermal comfort, *Ergonomics*, **14** (4), 457–468 (1971).
26. P. O. Fanger, *Conditions for Thermal Comfort—A Review*, Building Research Establishment, DoE, CIBComm., W45-Symp., Sept. 1972.
27. P. O. Fanger, *Thermal Comfort*, McGraw-Hill, 1972.
28. J. F. Nicol and M. A. Humphreys, *Thermal Comfort as Part of a Self-regulatory System*, Building Research Establishment, DoE, CIB Comm., W45-Symp., Sept. 1972.
29. D. McK. Kerslake, *The Stress of Hot Environments*, Cambridge University Press, 1972.
30. B. Givoni, *Man, Climate and Architecture*, Elsevier, Arch. Science Series, 1969.
31. T. Benzinger and G. Kitzinger, A 4π radiometer, *Rev. scient. Instrum.* **31**, 599–604 (1950).
32. D. L. Baun and P. E. McNall, A radiometer for environmental applications, *Trans. ASHRAE* **75**, Pt. 1 (1969).
33. O. M. Lidwell and D. P. Wyon, A rapid response radiometer for the estimation of mean radiant temperature, *J. scient. Instrum.*, Series 2, **1**, 534–538 (1968).
34. H. M. Vernon, The globe thermometer, *Proc. Instn. Heat. Vent. Engrs.*, **39**, 100 (1932).
35. E. A. Doebelin, *Measurement Systems: Application and Design*, McGraw-Hill, 1966.
36. D. M. McIntyre, A guide to thermal comfort, *Appl. Ergon.* 66–72, June 1973.
37. G. W. Brundrett, An introduction to thermal comfort in buildings, *Heat. Vent. Engr.*, 365–372 (1974).
38. M. A. Humphreys, *Environmental Temperature and Comfort*, BRE Current papers, CP 80/74, Building Research Station, 1974.

217

39. T. L. Madsen, *Thermal Environmental Parameters and their Measurement*, Building Research Establishment, DoE, CIB Comm., W45-Symp., Sept. 1972.

40. F. Schwartz, *The Comfortmeter*, Building Research Establishment, DoE, CIB Comm., W45-Symp., Sept. 1972.

41. P. W. O'Callaghan and S. D. Probert, Thermal properties of clothing fabrics, *Bldg. Serv. Engr.* **44**, 71–79 (1976).

42. S. T. Henderson and A. M. Marsden, *Lamps and Lighting*, 2nd edn., SI units, Edward Arnold, London, 1972.

43. H. C. Hottel and J. R. Howard, *New Energy Technology—Some Facts and Assessments*, MIT Press, 1971.

44. B. J. Brinkworth, *Solar Energy for Man*, Compton Press, England, 1972.

45. W. P. Jones, *Air Conditioning Engineering*, SI edn., Edward Arnold, London, 1974.

46. J. F. Van Straaten, *Thermal Performance of Buildings*, Elsevier, 1967.

47. H. K. Bourne, *Mans Effect on Climate*, DES, UKSM report No. 70/20, 1970.

48. R. H. Pennington, *Introductory Computer Methods and Numerical Analysis*, Macmillan, New York, 1965.

49. H. C. Hottel, Radiant heat transmission, *Mech. Engng.* **52** (1930).

50. K. C. Rockey, H. R. Evans, D. W. Griffiths and D. A. Nethercot, *The Finite Element Method*, Crosby Lockwood and Staples, London, 1975.

51. L. P. Huelsman, *Basic Circuit Theory with Digital Computations*, Prentice-Hall, New Jersey, USA, Electrical Engineering Services, 1972.

52. E. Danter, Periodic heat flow characteristics of simple walls and roofs, *J. Instr. Heat. Vent. Engrs.* 136–146 (1960).

53. C. O. Mackey and L. T. Wright, Periodic heat flow—homogeneous walls or roofs, heating, piping and air conditioning, *ASHVE Jl.* 546–555 (1944).

54. V. Paschkis and M. D. Baker, A method for determining unsteady-state heat transfer by means of an electrical analogy, *Trans. Am. Soc. mech. Engrs.* **64**, 105–112 (1942).

55. D. I. Lawson and J. H. McGuire, The solution of transient heat flow problems by analogous electrical networks, *Proc. (A) Instn. Mech. Engrs.* **167** (3) 275–287 (1953).

56. R. G. Nevins, Psychrometrics and modern comfort, paper presented at Joint ASHRAE–ASME Meeting, Nov. 1961.

57. P. W. O'Callaghan and S. D. Probert, Real area of contact between a rough surface and a softer, optically flat surface, *J. mech. Engng. Sci.* **12** (4) 256–267 (1970).

58. P. W. O'Callaghan and S. D. Probert, Thermal resistance and directional index between smooth non-wavy surfaces, *J. mech. Engng. Sci.* **16** (1) 41–55 (1974).

59. F. R. Al-Astrabadi, P. W. O'Callaghan, A. M. Jones and S. D. Probert, Thermal contact conductance correlations for stacks of thin layers in high vacuum, *J. Heat Transfer*, **99C** (1) 139–142 (1977).

60. R. Siegel and J. R. Howell, *Thermal Radiation Heat Transfer*, McGraw-Hill, 1972.

61. J. F. Malloy, *Thermal Insulation*, Van Nostrand Reinhold, Environmental Engineering Series, New York, 1969.

62. S. D. Probert and H. D. Malde, Free convection in open-ended rectangular cavities—inclination dependence, *J. mech. Engng. Sci.* **14**, 78–82, (1972).

63. P. W. O'Callaghan, S. D. Probert and G. J. Newbert, Velocity and temperature distributions in and around cold air jets issuing from linear slot vents into relatively warm air, *J. mech. Engng. Sci.* **17** (3), 139–149 (1975).

64. W. M. Kays and A. L. London, *Compact Heat Exchangers*, 2nd edn., McGraw-Hill, London, 1964.

65. J. E. Coppage and A. L. London, The periodic flow regenerator—a summary of design procedure, *Trans. Am. Soc. mech. Engrs.* **75**, 779–787 (1953).

66. D. Chisholm, *The Heat Pipe*, M & B Technical Library, Mills & Boon, London, 1971.

67. P. W. O'Callaghan, A. M. Jones and S. D. Probert, Review of developments in thermal rectification and storage, paper no. C111/76, presented at *6th Thermodynamics and Fluid Mechanics Convention, Institution of Mechanical Engineers, University of Durham, April,* 1976.

68. G. F. C. Rogers and Y. R. Mayhew, *Engineering Thermodynamics, Work and Heat Transfer*, 2nd edn., Longmans, London, 1970.

69. H. M. Power, Utilisation of wind power, paper presented at conference on *The Selection of Energy Resources for Housing, University of Salford,* September 1976.

70. J. M. Savins, Wind power, *Astronaut. Aeronaut.* Nov. 1975.

71. G. L. Dugger, Ocean thermal efficiency conversion, *Astronaut. Aernaut.* Nov. 1975.

72. A. L. Johnson, Biomass energy, *Astronaut. Aeronaut.* Nov. 1975.

73. M. Wolf, Photovoltaic power, *Astronaut. Aeronaut.* Nov. 1975.

74. P. Threlkeld, *Thermal Environmental Engineering*, 2nd edn., Prentice-Hall, New York, 1970.

75. T. Eiloart, Man with energy answers, *New Scient.* 782–786, Dec. 1973.

76. R. L. Fullman, Energy storage by flywheels, 10th *Intersociety Energy Conference, IEEE, New York*, 1975.

77. K. W. Li, Compressed air storage in gas turbine systems, *J. Engng. Power* 640–644 (1975).

78. P. W. O'Callaghan and S. D. Probert, Thermal accumulators, *Appl. Energy* **3**, 51–64 (1977).

79. E. G. Kovach (ed.), *Thermal Energy Storage*, Report of NATO Science Comm. Conf., Turnberry, Scotland, March 1976.

80. R. M. Diamant, *Total Energy*, Pergamon Press, Oxford, 1970.

81. P. W. O'Callaghan and S. D. Probert, Energy management, *Appl. Energy* **2**, (1977).
82. G. Meckler, Energy integrated design and its effect on building energy requirements, *ASHRAE Jl.* 54–62, Nov. 1965.
83. R. Hickman, Heat recovery in air conditioning, *Light Ltg.* 352–355, Oct. 1971.
84. Editorial announcement, The international system of units, *Int. J. Heat Mass Transfer* **9**, 837–844 (1966).

Additional Bibliography

Angrist, S. W., *Direct Energy Conversion*, Allyn & Bacon Inc., 2nd edn., New York, 1971.

Applied Solar Energy Research, 2nd edn., Pergamon, 1959.

Ashley, H., *et al.*, *Energy and the Environment: Risk Benefit Approach*, Pergamon Press, New York, 1976.

ASME, *Proceedings of the Conference Intersociety Energy Conversion, San Francisco, August 1974*, ASME, New York, 1974.

Berridge, G. L. C., *Heat Pumps: Key References in the Literature*, Brainchild Information Services, 1975.

Billington, N. S., *Building Physics: Heat*, Pergamon Press, Oxford, 1967.

Blair, I. M., Jones, B. D. and Van, A. J., *Aspects of Energy Conversion, Proceedings of the UK SRC Summer School, Oxford, England*, Pergamon Press, 1975.

Boer, K. W. (ed.), *Proceedings of the Joint Conference American Sect. Solar Energy Society and Solar Energy Society, Canada: Sharing the Sun—Solar Technology in the Seventies, Winnipeg, Canada, August 1976*, Pergamon Press, New York, 1976.

Boyen, J. L. *Practical Heat Recovery*, New York, Wiley, 1975.

Bronstead, J. N., *Principles and Problems of Energetics*, Interscience, 1955.

Chang, S. S. L., *Energy Conversion*, Prentice-Hall, London, 1963.

Courtney, R. G. (ed.), *Proceedings of the Conference Energy Conservation in the Built Environment*, Building Research Station, Construction Press Ltd., 1976.

Dunn, R. and Reay, D. A., *Heat Pipes*, Pergamon Press, Oxford, 1976.

Electrochemical Society, *Proceedings of the International Symposium Solar Energy*, Washington DC, *May* 1976, *Princeton, New Jersey*, Electrochemical Society, 1976.

Fisher, J. C., *Energy Crisis in Perspective*, John Wiley, New York, 1974.

Great Britain Directorate of Building Development, *Solar Heat and the Overheating of Buildings*, HMSO, 1975.

Grawshaw, T. (comp.), *Alternative Sources of Energy*, GEC bib 154, GEC Power Systems Library, 1976.

Harrah, B. and Harrah, D., *Alternative Sources of Energy*, Methuen, NJ, Scarecrow Press, 1975.

Hickcock, F., *Handbook of Solar and Wind Energy: Cahners Special Report*, Boston, Mass., Cahners Books, 1976.

Holloman, J. H. and Grenon, M., *Energy Research and Development*, Cambridge, Mass., Ballinger (Wiley, London), 1975.

Hunt, S. E., *Fission, Fusion and the Energy Crisis*, Pergamon Press, Oxford, 1974.

IEEE, *Proceedings of the Conference Intersociety Energy Conversion, University of Delaware, Newark, August 1975*, IEEE, New York, 1975.

Ion, D. C., *Availability of World Energy Resources*, London, Graham & Trotman, 1975.

Institution of Mechanical Engineering, *Energy Recovery in Process Plants*, Proc. Conf. Instn. Mech. Engrs. and DoE, January 1975, 1976.

International Solar Energy Society, *Solar Energy: A UK Assessment*, London, UK Section, ISEC, 1976.

Karam, R. A. and Morgan, K. Z., *Energy and Environment—Cost-Benefit Analysis*, Pergamon Press, New York, 1976.

Karam, R. A. and Morgan, K. Z., *Environmental Impact of Nuclear Power Plants*, Pergamon Press, New York, 1976.

Kovach, E. G., *Technology of Efficient Energy Utilisation, Report of NATO Science Comm. Conference, Les Arcs, France, 1973*, Pergamon Press, Oxford, 1974.

Kreider, J. F. and Kreith, F., *Solar Heating and Cooling: Engineering, Practical Design and Economics*, McGraw-Hill, New York, 1975.

Larson, E., *New Sources of Energy and Power*, London, Muller, 1976.

Leach, G., *Energy and Food Production*, IIED, 1975.

Lindsey, R. B. (ed.), *Energy: Historical Development of the Concept*, Stroudsberg, Penns.; Dowden, Hutchinson & Ross, Wiley, London (Halsted Press), 1975.

Lovins, A. B., *World Energy Strategies: Facts, Issues and Options*, Friends of the Earth, 1975.

McDougall, A., *Fuel Cells*, Macmillan, London, 1976.

McMullan, J. T., *et al.*, *Energy Resources and Supply*, Wiley Interscience, London, 1976.

McVeigh, J. C., *Sun Power*, Pergamon Press, Oxford, 1977.

Meinel, A. B., and Meinel, M. P., *Applied Solar Energy*, Reading, Mass., Addison-Wesley, 1976.

Messel, H. and Butler, S. T., *Solar Energy*, Pergamon Press, Oxford, 1975.

Murray, R. L., *Nuclear Energy*, Pergamon Press, New York, 1975.

National Aeronautics Space Administration, *Proceedings of the International Solar Energy Conference, NASA, Goddard Space Flight Center, Maryland*, Pergamon, New York, 1971.

Portola Inst., *Energy Primer : Solar, Water, Wind and Biofuels*, Menlo Park, Calif., Prism Press, Dorchester, Dorset, 1974.

Preist, J., *Energy for a Technological Society : Principles, Problems, Alternatives*, Addison-Wesley, Reading, Mass., and London, 1975.

Reay, D. A., *Industrial Energy Conservation*, Pergamon Press, Oxford, 1977.

Reynolds, W. C., *Energy : From Nature to Man*, McGraw-Hill, New York, 1974.

Simon, A. L., *Energy Resources*, Pergamon Press, New York, 1975.

Smith, C. B., *Efficient Electricity Use*, Pergamon Press, Oxford, 1976.

Sporn, P., *Energy in an Age of Limited Availability and Delimited Applicability*, Pergamon Press, New York, 1976.

Stoeker, W. F., *Design of Thermal Systems*, McGraw-Hill, 1971.

Sumner, J. A., *Domestic Heat Pumps*, Dorchester, Prism Press, 1976.

Sutton, G. W., *Direct Energy Conversion*, McGraw-Hill, 1966.

Szekely, J. (ed.), *The Steel Industry and the Energy Crisis*, Marcel Dekker, 1975.

Voegeli, H. E. and Tarrant, J. J., *Survival 2001 : Scenario from the Future*, Van Nostrand and Reinhold, New York and London, 1975.

Index

223